ROUTLEDGE LIBRARY EDITIONS: HISTORY AND PHILOSOPHY OF SCIENCE

EVOLUTIONARY THEORY AND CHRISTIAN BELIEF

EVOLUTIONARY THEORY AND CHRISTIAN BELIEF

The Unresolved Conflict

DAVID LACK

Volume 21

Routledge
Taylor & Francis Group
LONDON AND NEW YORK

First published in 1957

This edition first published in 2009 by
Routledge
2 Park Square, Milton Park, Abingdon, Oxon, OX14 4RN

Simultaneously published in the USA and Canada
by Routledge
270 Madison Avenue, New York, NY 10016

Routledge is an imprint of the Taylor & Francis Group, an informa business

British Library Cataloguing in Publication Data
A catalogue record for this book is available from the British Library

Library of Congress Cataloging in Publication Data
A catalog record for this book has been requested

ISBN 10: 0-415-42029-6 (Set)
ISBN 10: 0-415-47490-6 (Volume 21)

ISBN 13: 978-0-415-42029-7 (Set)
ISBN 13: 978-0-415-47490-0 (Volume 21)

Publisher's Note
The Publisher has gone to great lengths to ensure the quality of this reprint but points out that some imperfections in the original copies may be apparent.

Disclaimer
The Publishers have made every effort to trace copyright holders and would welcome correspondence from those they have been unable to contact.

"But then arises the doubt, can the mind of man, which has, as I fully believe, been developed from a mind as low as that possessed by the lowest animal, be trusted when it draws such grand conclusions?" (C. Darwin, 'Autobiography', 1876)

This drawing, without title, signature or date, but presumably Victorian, is in the Department of Zoology, Leicester

EVOLUTIONARY THEORY AND CHRISTIAN BELIEF

The Unresolved Conflict

DAVID LACK

METHUEN & CO LTD

36 Essex Street, Strand

London

First published in 1957

Catalogue Number 5984/U
Printed in Great Britain by
Western Printing Services Ltd, Bristol

Contents

Preface

My aim in this book is to discuss a live and highly controversial issue in simple and I hope fair terms. Various writers, both Christian and agnostic, have claimed that the dispute is over, but this, I suggest, is because they have not accepted the full implications of evolution by natural selection, or alternatively of Christianity. The basic conflict is unresolved, yet it is important for everyone to study man's origin and nature. I have therefore used few technical terms, and have tried to give the arguments so that they can be appreciated, if not accepted, by both Christian and agnostics, while rejecting false claims by both sides.

For convenience, and despite its Victorian flavour, I have often used the term 'Darwinism', in a biological sense for the evolution of animals from pre-existing forms by the natural selection of hereditary variations, and also in a philosophical sense for the view that man has evolved from animals wholly by such means. (But 'Darwinist' is used more loosely, for any prominent supporter of Darwin's views in the early days.) The problems to be discussed are immensely difficult, ranging over the borderland between biology, theology and philosophy, and I have been trained only in biology, so that I am haunted by the words of an earlier naturalist who ventured into this field that 'many . . . have too rashly charged the troops of error, and remain as trophies unto the enemies of truth' (Sir Thomas Browne, *Religio Medici*, 1643). My aim is to make what is known clear, not to penetrate unexplored depths; also to keep within the limits of my title, and not to follow any topic in its wider implications.

My debt to previous writers is great; their books are cited in the text by title, except that the one word 'Life' is used where reference is to the standard Life and Letters of the person mentioned. Page references to quotations are given in Appendix I. I am also extremely grateful to the following critics, Roman Catholic, Anglican, Quaker, and agnostic, biologist, philosopher, priest and layman: Dr. A. J. Cain, Dr. A. C. Crombie, Mr. R. E. Moreau, Mr. D. Neylan, Fr. Pierce, C.R., Fr. J. L. Russell, S.J., Dr. W. H. Thorpe, F.R.S., Dr. B. Towers and Rev. J. M. Wilson, M.D. Their vigorous and stimulating criticisms showed me many errors and helped me to appreciate their divergent viewpoints. Finally, I am most grateful to Professor Moon and the Department of Zoology, University College, Leicester, for the loan of the pen-and-ink drawing, unsigned and undated, found in one of their corridors, which is reproduced as the frontispiece of this book; but I am responsible for entitling it, I hope appropriately, with an extract from Darwin's autobiography (published in his *Life and Letters*).

1

Historical Introduction

'Grant that the building now to be erected on this spot may foster the progress of those sciences which reveal to us the wonders of Thy creative powers. And do Thou, by Thy heavenly grace, cause the knowledge thus imparted to fill us with the apprehension of Thy greatness, Thy wisdom and Thy love.' With this prayer, specially composed by the Professor of Medicine, the foundation stone was laid in 1855 for the Oxford Museum of Science; but the implied unity between religion and science was broken at the first important meeting to be held there, that of the British Association for the Advancement of Science in 1860, when Samuel Wilberforce and T. H. Huxley held their famous debate on evolution (H. M. and D. Vernon, *History of the Oxford Museum*, 1909).

In 1858, the Linnean Society of London heard the joint contributions by Charles Darwin and Alfred Russel Wallace on evolution by natural selection, and Darwin's *The Origin of Species* appeared in the following year, the first edition being sold out at once. But in popular imagination the conflict was truly joined in June 1860, when the Bishop of Oxford was put up to oppose Darwin's new theory in public debate. T. H. Huxley, though attending the British Association meeting, had meant to stay away from the debate in question, but was pressed into coming at the last moment. What was said on that occasion has been reported, in confident quotation marks, in many books, but no verbatim account was kept, and different persons recollected both the words and the emphasis differently. Even Huxley himself could not recall what he said. It is agreed that the Bishop made a polished and witty but superficial speech, ending with an objectionable allusion to human descent from an ape. The historians of science often imply that he was personally abusive to Huxley, and this version goes unchallenged, since the meeting was mentioned with suspicious brevity in the three-volume life of Wilberforce by his son. But such rudeness does not seem in keeping with Wilberforce's character, and Professor Farrar, who was present, thought the words used were 'flippant and unscientific rather than insolent, vulgar or personal':' "If any one were to be willing to trace his descent through an ape as his *grandfather*, would he be willing to trace his descent similarly on the side of his *grandmother*?" ', thus with misplaced humour trying to arouse a sentimental objection to the idea of woman being so degraded. Even so, it seems inexcusable, and Huxley, murmuring to his neighbour 'The Lord hath delivered him into mine hands' (strange comment from an agnostic), gave a plain and honest account of Darwin's scientific views and ended to the effect that he would rather have a monkey for his grandfather than one who used great gifts to stifle truth. (Huxley, *Life*, 1903.)

One cannot say what might have happened if the Bishop of Oxford had been more earnest and Huxley had replied in the spirit of the New Testament not the Old, but probably the debate of the next thirty years would have been no less violent. For after a short time, it was not concerned solely with the truth of a particular theory; biologists regarded it as a fight for the freedom of scientific enquiry against religious dogma and prejudice, of truth against authority, and theologians as a fight to preserve the spiritual relationship of man to his Creator in a rising tide of materialism.

Before continuing, it is worth glancing at the two antagonists in this famous encounter. One pictures Huxley as in his later portraits, the eminent and forceful Victorian, but he was only thirty-six at the time and as yet unpractised in debate. A great anatomist, he devoted the rest of his life to what he held to be his public service; battling in lectures and essays on behalf of Darwinism and in later years against Christian beliefs, in a style that was vigorous and incisive because he so fervently believed what he said; and labouring with great industry, force and tact as secretary of the Royal Society and as a member of ten Royal Commissions and of the London School Board. Various views of his will find place in later chapters, and his general attitude may be summed up in two quotations. In 1856, as his wife lay in the next room awaiting the birth of their first child, he listed among his aims for the future: 'To smite all humbugs, however big; to give a nobler tone to science'. In 1860, just after that same boy had died, he wrote in a long and moving personal letter to Charles Kingsley: 'Sit down before fact as a little child, be prepared to give up every preconceived notion, follow humbly wherever and to whatever abysses nature leads, or you shall learn nothing.' (*Life*, 1903.)

Samuel Wilberforce was also a great man. Like Huxley he had rare personal charm, energy and zeal, and could hold an audience of Lords or workingmen, being far more than the glib speaker of the historians of science, though his facility led him astray at times. He effected much-needed reforms in episcopal administration, introduced parochial missions, founded the theological college at Cuddesden and supported in his diocese the first Anglican order for men since the Reformation. Mr. Gladstone wrote of him in a letter to Queen Victoria that he believed 'there does not live the man in any of the three kingdoms of your Majesty who has by his own indefatigable and unmeasured labours given such a powerful impulse . . . to the religious life of the country' (C. P. S. Clarke, *The Oxford Movement and After*, 1932). He loved nature, and in lighter vein is credited with the lines on a Cassowary on the plains of Timbuktoo, that ate a missionary, coat and bands and hymnbook too. He was forty-five at the time of his encounter with Huxley.

In the same year, 1860, Wilberforce reviewed *The Origin of Species* in the *Quarterly Review*, helped, it is thought, by Sir Richard Owen, England's leading but by then very conservative anatomist. The review was adverse and prejudiced, though not so unpleasant in tone as that by Owen himself for the *Edinburgh Review* three months earlier. The style was brilliant, but the scientific arguments were confused, apart from a few shrewd points, such as that Darwin supposed the spots of the young Blackbird useless, but Wilberforce (with more justification) that they were protectively coloured. Wilberforce accepted the idea of natural selection, but forcibly argued that it could not account for man's peculiar moral and spiritual condition (a view later supported by some of the leading Darwinists). The Bishop was at his best in setting forth what he held to be the Christian attitude towards scientific truth: 'We have no sympathy with those who object to any facts or alleged facts in nature, or to any inference logically deduced from them, because they believe them to contradict what it appears to them is taught by Revelation. . . . To oppose facts in the natural world because they seem to oppose Revelation . . . is . . . but another form of the ever-ready feebleminded dishonesty of lying for God, and trying by fraud or falsehood to do the work of the God of truth. It is with another and a nobler spirit that the true believer walks amongst the works of nature. The words graven on the everlasting rocks are the words of God, and they are graven by His hand.' . . . They cannot 'contradict His Word written in His book. . . . There may be to man difficulty in reconciling all the utterances of the two voices. But what of that? He has learned already that here he knows only in part, and that the day of reconciling all apparent contradictions between what must agree is nigh at hand.' This passage suggests sincere conviction, and it may be regretted that the controversy was not to be continued in a like spirit.

Soon after a great man has died there is a tendency to denigrate his achievement, and Darwin's turn came, though later than is usual, his reputation being lowest from about 1920 to 1930. Unfortunately, three widely read histories of science, by Nordenskiöld, Rádl and Singer, were written in this period, so that the general reader may get a false impression of Darwin's contribution. It is true, as these writers pointed out, that others had advocated animal evolution before Darwin, but they carried little weight; and their more important criticism, that Darwin's theory of natural selection was later shown to be of little value and mainly disproved, is definitely wrong. Of Darwin's greatness and originality there is no serious doubt. It is apparent in the majestic ordering of facts and arguments in the pages of *The Origin of Species*. Moreover he anticipated many of the biological objections that were to be raised to his theory and admitted the main gaps in his evidence, gaps that are to his credit rather than discredit, since he was able to see beyond them to his grand design. He was ignorant of the true nature of heredity, but so were all biologists of the time; and his theory of natural selection has been re-established as the corner-stone of modern evolutionary theory. Only an intellectual giant could have initiated so great a turmoil. He himself took no part in the resulting controversies when they were outside the field of biology, except for a few comments in private letters. He therefore finds little further place in this book, though it may be added that he gradually and unobtrusively lost his belief in Christianity.

Darwinism conflicted, or was thought to conflict, with Christian belief in a number of different but interconnected ways. It contradicted the account of creation in Genesis and hence challenged the truth of the Bible. It undermined what was at that time the most popular rational argument for the existence of God, from the presence of design and apparent purpose in the animal body, claiming instead that such adaptations had come into existence by wholly natural means, of a seemingly random and rather bloodthirsty kind. It ran counter to the historical occurrence of the Fall, suggesting that man had risen from the beasts not fallen from a state of blessedness, thus questioning whether man was in a state of sin. Further, if man's higher capacities had been evolved by natural means from those of animals, they might have no ultimate value or significance. Darwinism thus challenged, or was believed to challenge, the fundamental view that man was created by God in His image and stood in a special relationship towards Him.

These issues were not raised in *The Origin of Species*, but it was obvious that they soon would be raised, so that many theologians feared that Christian belief would be undermined. Before criticising the churchmen, however, it is as well to recall that various biologists, notably Sir Richard Owen, were equally hostile, and that in 1864 the President of the Royal Society, in giving the Copley Medal (its highest award) to Darwin, announced that 'speaking generally and collectively, we [the Council] have expressly omitted it [the theory of evolution] from the grounds of our award.' (Huxley, *Life*, 1903.) It is almost impossible for us, a century later, to think ourselves back into the position of people of the time, and only too easy to criticize at this safer distance statements made under great emotional stress. Moreover, later events showed that churchmen were right in supposing that Darwinism would lead many away from Christianity, though some might argue that this was due less to Darwinism itself than to the mistaken ways in which theologians tried to refute its claims. Further, the two leading propagandists for evolution, T. H. Huxley in Britain and Ernst Haeckel in Germany, were notoriously opposed to Christianity on other grounds, and wrote against miracles and dogma. It was, of course, T. H. Huxley who coined the term 'agnostic'. Finally, the Church was at the same time strongly attacked from another direction, since the applications of historical and textual criticisms to the Bible were destroying belief in its complete accuracy and thus challenging its authority; but this battle the orthodox theologians appear to have fought much more effectively than that against the biologists.

As just one illustration of the intense excitement aroused by Darwinism, both of England's great Prime Ministers made contributions, very different in tone, but equally in keeping with their respective characters. The depressing point about Disraeli's shallow witticism is that it was not recognized as such. Addressing the Oxford Diocesan Society in the Sheldonian Theatre in 1864, he announced: 'I am not prepared to say that the lecture-room is more scientific than the Church (cheers). What is the question now placed before society with a glib assurance the most astounding? The question is this—Is man an ape or an angel? (loud laughter). My lord, I am on the side of the angels (laughter and cheering). I repudiate with indignation and abomination those views (cheers).' (B. Disraeli, *Church Policy*, 1864.) We still smile, yet for the true Christian as much for the scientist it was a ghastly frivolity, especially coming from the leader of the nation. Twenty years later, W. E. Gladstone found time in his busy career to demonstrate (in the *Nineteenth Century* for 1886) that the order of creation in Genesis was that accepted by scientists, his laboured erudition being almost as misdirected as in his earlier attempt to wed the Bible with Homer.

There is no need to repeat here either the ill-founded attacks of churchmen on Darwinism, which have often been recalled, or the ill-founded attacks of Darwinists on the Church, which are usually forgotten. It is enough to say that, while many churchmen, both Roman Catholic and Protestant, showed a lamentable ignorance of the findings and the principles of biology, the same could be said of various Darwinists in relation to theology. Mixed up with the truth, there were ignorant, unjustifiable, absurd and violent assertions on both sides, and it is perhaps through the spirit of the age that we remember the arrogance of conservative theologians rather than of revolutionary Darwinists.

It is more profitable to record that a few churchmen, including some of the most distinguished, welcomed Darwin's theory from the start. F. J. A. Hort wrote in a letter in March, 1860: 'Have you read Darwin? . . . In spite of difficulties, I am inclined to think it unanswerable.' (*Life*, 1896.) R. W. Church, later Dean of St. Paul's, also admired it (*Life*, 1894). Charles Kingsley both praised the book and wrote sympathetic letters to Darwin and Huxley (*Life*, 1877). Several other Anglicans felt similarly, while on the Roman Catholic side, Cardinal Newman wrote in a private notebook of 1863 'It is strange that monkeys should be so like men with no historical connection between them. . . . I will either go the whole hog with Darwin or, dispensing with time and history altogether, hold not only the theory of distinct species but also of the creation of fossil-bearing rocks.' (Quoted by H. Johnson, *Dublin Rev.*, 195 : 46, 1934.)

If men with such views had not kept their approval to private letters, they might have helped to bridge the widening gap between science and religion. Hort 'burned to speak openly' and had meant to review *The Origin of Species*, but held back. Kingsley mentioned in 1863 that he would write about it, 'but I am not going to reach into fruit this seven years'; he never did so, except to affirm the validity of both the scientific and Christian standpoints in a lecture to theology students in 1871 (*Scientific Lectures and Essays*, 1880). Cardinal Newman explained that his silence was due to various difficulties. 'One of the greatest is this, that at the moment it is so difficult to say precisely what it is that is to be encountered and overthrown,' and therefore that 'it seemed to be a time in which Christians had a call to be patient' (*Apologia pro Vita sua*, 1864). With this view many would agree, while regretting that less able persons showed less restraint.

So far as I am aware, the first substantial and sympathetic appraisal of animal evolution by an ecclesiastic was that in the *Dublin Review* for 1871 by the Roman Catholic Canon Hedley, who maintained that 'it is not contrary to Faith to suppose that all living things, up to man exclusively, were evolved by natural law out of minute life-germs primarily created, or even out of inorganic matter. On the other hand, it is heretical to deny the separate and special creation of the human soul; and to question the immediate . . . formation by God of the bodies of Adam and Eve . . . is, at least, rash, and, perhaps, proximate to heresy.' Sixteen years later, there came on the Anglican side a fuller and more sympathetic judgment, with some illuminating comments, by the Reverend Aubrey Moore (*repr.* in *Science and Faith*, 1889). Moore held that the human body was evolved by natural means from other animals, but that his soul came by divine gift.

A similar distinction was made by two of the leading biologists of the time. A. R. Wallace, a great naturalist, formulated the theory of natural selection independently of Darwin, while he was in the Malay Archipelago, and he later ascribed more importance to it as the agent of evolution than did Darwin. He held, however, that though the human body was evolved by natural means, some sort of intervention or intelligent agent was needed to account for man's higher nature, and notably for his mathematical and artistic faculties (*Darwinism*, 1889). St. George Mivart, a skilled anatomist and a Roman Catholic, fully accepted evolution as against special creation, but supposed natural selection to be of very minor importance, and also held that the human soul, including that of the first man, was 'absolutely created in the strict and primary sense of the word

. . . by a supernatural act' (*On the Genesis of Species*, 1871). For such a compromise, Wallace and Mivart were scorned in the *Quarterly Review* for 1871 by T. H. Huxley, whose grandson even described Wallace as 'unorthodox' in believing a supernatural intervention needed for man's soul (*Dict. Nat. Biogr.*, 1912–21). On each side extreme views found favour, though few agreed with Samuel Butler, who violently attacked both the Church and Darwinism, and hardly anyone heeded Dean Farrar 'that true science and true religion are twin sisters, each studying her own sacred book of God, and nothing but disaster has arisen from the petulant scorn of the one and the false fear and cruel tyrannies of the other' (*History of Interpretation*, 1886).

This summary will, I hope, show the historical background of the controversy, so far as Britain is concerned. Full information for Europe and America, and for all forms of Christianity, can be found in A. D. White's *History of the Warfare of Science with Theology in Christendom* (1896), and there are sympathetic summaries, on which I also drew, in: S. C. Carpenter, *Church and People, 1789–1889* (1933); L. E. Elliott-Binns, *Religion in the Victorian Era* (1936); W. L. Knox and A. R. Vidler, *The Development of Modern Catholicism* (1933); E. C. Messenger, *Evolution and Theology* (1931); and C. C. J. Webb, *A Study of Religious Thought in England from* 1850 (1933). The views of other writers, where important, will be discussed later under the subjects concerned.

The fury of the conflict subsided, and Darwin, once thought of as the arch-enemy of Christianity, was buried with honour in Westminster Abbey. His views were later attacked by various atheists, who held them to be incompatible with man's higher nature, and thereafter the dispute became many-sided, and milder. Indeed the modern tendency is to suppose that the conflict lies wholly in the past, though this seems largely because each side fails to appreciate or accept essential claims held by the other. Thus an uncharitable critic of recent books might suppose that evolutionary biologists think the conflict over because they have not properly grasped the arguments of theology or the complexity of man's higher nature; that Roman Catholics do so because, over-concerned with the letter of Genesis, they invoke a supernatural interference with natural laws at man's first appearance; and that Anglicans, together with various agnostics, do so because they advocate a pseudo-mystical means of evolution that is biologically unsound. This is an exaggerated and distorted statement, but it may help to reveal the underlying differences. I should add that there are theologians who accept the findings of biologists in the evolutionary field, while among evolutionary biologists of repute there are not only atheists and agnostics, but Unitarians, Quakers, Methodists, Presbyterians, Anglicans and Roman Catholics, who hold their divergent views with strong conviction and apparent integrity. This suggests, at least, that the problem is not simple.

2

The Fact of Evolution

Darwin's great book included two fundamental ideas, first that the many different kinds of animals on the earth were not specially and individually created, but have been modified by gradual changes from pre-existing forms over a huge length of time, and secondly that the main agent of evolution has been natural selection. The second of these ideas will be considered in Chapter 4. In regard to the first, the occurrence of evolution came to be accepted within a few years of the publication of *The Origin of Species* and is now generally agreed. Even today, a few Christians seem to consider it their duty to imply that serious doubt still exists, sometimes supporting this suggestion by quoting out of context an over-cautious statement by a biologist, perhaps as reported in the daily press. So let it be stated categorically that the evidence for the occurrence of animal evolution is overwhelming and that all serious students accept it. It may be added that this view is accepted by nearly all reputable religious writers, both Catholic and Protestant.

The evidence for evolution is so well known that there is no longer need to set it out in a book of this kind. The several strands include the similarities in structure between living animals, the gradual changes in structure revealed by the fossil record, the features of embryos, the facts of geographical distribution, and the hereditary changes known, in the first place, in domestic animals. It is true that no one has actually witnessed the transformation of one living species into another, but except in certain plants (polyploids), the hereditary changes involved are now known to be so numerous and complex that one could not expect them to be achieved by natural means in so short a time as one century. Important evidence now available for the occurrence of evolutionary change below the level of the species will be considered in Chapter 4.

Although the change from one species into another must normally occur gradually over many generations, each living species is usually demarcated sharply from others and does not intergrade with them in its characters. Darwin omitted from his book any serious discussion of how the gaps between species might arise. It has even been claimed that his book was misnamed, since it was concerned with gradual change and not with the separation of forms. But this is to overstress the point, since Darwin solved the greater problem, the origin of divergence. The lesser problem, the origin of discontinuity, was not solved until the last twenty-five years. Evidence has now been pieced together to show that, when two populations of one species become isolated from each other, they gradually acquire hereditary difference. If the isolation persists for a sufficiently large number of generations, they may become so different that, if they again meet, their hereditary characters do not mingle well in any hybrid offspring, which are at a disadvantage compared with either parent type. Under these circumstances, parents that breed with their own form will tend to leave more survivors than those which hybridize. As a result, natural selection will favour the evolution of psychological or physical barriers to interbreeding, and in time the two forms will keep entirely separate and new species will have arisen. At least in the larger animals, like birds and mammals, the isolation of populations that precedes the formation of new species is geographical; new species arise from races or subspecies living in separate areas. These new subspecies normally arise by changes occurring in whole populations, not in just a few individuals, and recent research has stressed the importance of the population as a unit of evolution, at least in higher animals. (The special cases of forms that reproduce asexually and of polyploid plants need not be discussed here, as they are not typical of the evolutionary process as a whole, for which see G. G. Simpson, *The Meaning of Evolution*, 1951; Th. Dobzhansky, *Genetics and the Origin of Species*, 1951; E. Mayr, *Systematics and the Origin of Species*, 1942.)

Darwinism is concerned with the evolution of living animals and plants from one or a few pre-existing forms, and the problem of how life itself originated is outside the scope of this book. It was formerly believed that fleas came from dust, flies from putrefaction, and even geese from barnacles, while after these crude ideas had been disproven, Pasteur had a long struggle to establish that micro-organisms likewise arise from other micro-organisms, and not by spontaneous generation from non-living matter. The usual belief among scientists today is that living matter arose from non-living matter under peculiar physical and chemical conditions prevailing far back in the earth's past, and not since repeated. But this is only a supposition, and the view that new life is still arising around us from non-living matter, but at the submicroscopic level, cannot yet be ruled out. Because the problem is unsolved, certain religious writers have postulated that the origin of life required a miracle, but this view may be whole-heartedly condemned, not only on scientific but also on theological grounds, for there is a proper reluctance to postulate special divine interferences with the natural organization of the universe.

In Darwin's time, the claim that man had evolved from other animals was based on anatomical similarities, as well shown, for instance, in T. H. Huxley's short book, *Evidence as to Man's Place in Nature*, of 1863. Since then, some important fossils have been found, and the following summary is based on the review of them by W. E. Le Gros Clark (*The Fossil Evidence for Human Evolution*, 1955). Man belongs to the primates, the order of mammals which also includes apes, monkeys and lemurs. Among living primates, he is most closely related to the anthropoid apes, such as the Gorilla and Chimpanzee, though not as closely as once thought, and their respective ancestors probably diverged from a common stock in the Miocene period, say twenty-five million years ago. But there are as yet few relevant fossils on which to form an opinion.

The earliest known fossils which are definitely on the human line of evolution, and away from that of the anthropoid apes, have been found during the last thirty years in caves and fissures in South Africa. They date from very early in the Pleistocene period, between half a million and a million years ago. Formerly it was thought that several species were involved, but it is now agreed that there is only one, named *Australopithecus transvaalensis*. This is so intermediate in its characters between modern man and higher ape as to justify the hackneyed term 'missing link'. The skeleton of a modern man differs from that of a higher ape especially in three features, the much larger brain-case, the smaller and simpler teeth, and the fully erect posture, the last being revealed by the structure of the limbs and pelvis and by the way in which the skull is set on the spine. The limbs and pelvis of *Australopithecus* show that it had evolved a long way towards an erect posture, the skull and teeth had some human and some primitive features, but the jaw was massive and protruding, while the brain-case was much smaller than in modern man, and only slightly larger than in an ape. Provisionally at least, *Australopithecus* seems best regarded as a pre-human phase in man's ancestry, rather than as a true man. The use of tools is uncertain, an earlier claim to that effect being now discounted, but pebble tools in river deposits are probably contemporary with the later skeletons, while many skulls of baboons found with *Australopithecus* were fractured as if with an instrument.

The gap between *Australopithecus* and modern man is bridged by the fossil *Pithecanthropus erectus* of Java, found in 1891, while much further material was discovered between 1936 and 1941. Between 1926 and 1939 similar skeletons were found near Pekin. The latter were named *Sinanthropus pekinensis*, but are better termed *Pithecanthropus pekinensis*, to show their close affinity to the Java form. Indeed, some authors (e.g. Mayr, *Cold Spring Harbor Symposia*, 15 : 109–18, 1950) have regarded Pekin and Java man as belonging to the same species, implying that they could freely interbreed. The brain-case of *Pithecanthropus* was much larger than in any known ape or in *Australopithecus*, and though it was smaller than in modern man, it was within the known range for our species. The posture was erect, and the teeth were of human type except for rather large canines, but there were heavy brow-ridges.

Simple tools of quartz flakes and chipped bone, also fires, were associated with the Pekin remains. It is therefore reasonable to classify *Pithecanthropus* as a man. Both the Java and Pekin specimens date from early in the Pleistocene, later than *Australopithecus*, but before the period from which our own species, *Homo sapiens*, is known. Both in time and in anatomy, *Pithecanthropus* could be ancestral to ourselves.

Another fossil, the Piltdown skull, need not detain us, having been shown to be a clever fake. This demonstration in no way invalidates the claim that man evolved from ape-like ancestors.

Skeletons certainly referable to *Homo sapiens* have been traced back to the warm period before the last glaciation (or ice age), some 50,000 years ago. If, as is probable, some skull-bones found at Swanscombe in Kent belong to true man, the date can be pushed back twice as far, to the warm period before the third glaciation, and a more dubious mandible from Heidelberg might double this period again.

The remaining human fossils, classified as *Homo neanderthalensis*, are an offshoot that did not persist. Neanderthal man existed alongside *Homo sapiens* at the start of the last glacial period and differed from us in various primitive features, such as the large brow-ridges, the jaws and palate, and the slouching gait, but it is now considered that these features were secondary, and that Neanderthal man evolved from a form which walked fully upright. He used fire and simple tools. Skeletons from Mount Carmel in Palestine are intermediate in appearance between Neanderthal man and primitive *Homo sapiens*, and may represent either the transition between the two or hybrids between them.

To conclude, the fossils found since Darwin's time fully

support the conclusion of the early Darwinists that man evolved from primate stock, and the discovery of *Australopithecus* carries man's known ancestry back to between half a million and a million years ago. During this period the most remarkable change was in the capacity of the brain, which increased in size far more rapidly than has been found in any other series of fossil animals.

Julian Huxley has argued (in *The Uniqueness of Man*, 1941) that man's powerful intellect could have been evolved only in an animal as against a plant, in a multicellular animal, in one which sought actively for food and so had both bilateral symmetry and a definite head, in a highly differentiated animal, in a vertebrate and one which, moreover, maintained a constant temperature, in a social species, bearing only one young at a birth, thus allowing a long period of learning and experience before maturity, and finally in a land animal which had earlier been arboreal, thus having evolved a prehensile hand which was later freed for other uses. Huxley also stressed that man has features, especially of the intellect, which make him unique among animals, thus correcting the impression given by the early Darwinists, who with understandable enthusiasm, overstressed the animal features of man.

Evolution is a scientific theory, formulated to account for the nature of animal and plant life on the earth. But particularly in view of the fossil record, evolution may also be termed a fact. It is not a scientific fact in the sense that the speed of sound, or the chemical composition of the liver, or the instinctive responses of a young bird, are facts—facts which can be repeatedly observed and checked. The course of evolution has run in particular directions and not others. General trends have produced parallel resemblances, as between whales and fish, or hawks and owls, which look like each other because they have independently evolved similar adaptations. But each species of animal is different from all others, and in so far as evolution is concerned with particular rather than general events, it seems closer to a historical than a scientific fact.

3

The Truth of Genesis

That animals evolved over great ages of time, and men from animals, is contrary to a literal reading of the creation story in Genesis, and this provoked the earliest and most violent of the conflicts between Darwinism and the Church. In retrospect it seems surprising, for in some ways this was the simplest of the issues involved, and moreover it had already been raised more than twenty years before the appearance of *The Origin of Species*. In his *Principles of Geology* of 1830–33, Sir Charles Lyell established the geological succession of stratified rocks and fossils, thus showing the world to be far older than the accepted date for the Garden of Eden as calculated from numerical information in the Bible. Also, in 1836 Dean Buckland said in effect, though most circumspectly, that the findings of geology were incompatible with the manner and order of creation described in Genesis, all geological history being contained in the opening phrase "In the beginning" (*Geology and Mineralogy considered with reference to Natural Theology*, one of the *Bridgewater Treatises*). But Darwin's book controverted a more dramatic part of the story of Eden. Even so, this aspect of the conflict might not have loomed so large if the accuracy of many other parts of the Bible had not been violently attacked on historical and textual grounds by various German authorities, whose writings began to be known in England at about the same time as Darwinism.

So great was the alarm that the 'Oxford Declaration' of 1864 was issued and signed by numerous Anglican clergymen: 'We declare our firm belief that the Church of England and Ireland, in common with the whole Catholic Church, maintains without reserve or qualification, the Inspiration and Divine Authority of the whole Canonical Scriptures, as not only containing but being the Word of God' (*Life of E. B. Pusey*, 1893). And for Roman Catholics, Pope Leo XIII pronounced in his encyclical *Providentissimus Deus* of 1893 that 'all the books, which the Church receives as sacred and canonical, are written wholly and entirely, with all their parts, at the dictation of the Holy Ghost; and so far is it from being possible that any error can co-exist with inspiration, that inspiration not only is essentially incompatible with error, but excludes and rejects it as absolutely and necessarily as it is impossible that God Himself, the supreme Truth, can utter that which is not true' (quoted by W. Temple, *Nature, Man and God*, 1934).

Scientists may condemn these pronouncements as obscurantist, but their aim, at least in part, was to restrain alarm and preserve the faith among ordinary Christians until the new research could be tested and assimilated. Much, also, depends on their interpretation, and 'the Pope goes on to say that the sacred writers consequently speak of the constitution of the visible world and of its phenomena as the men of their time spoke and in language intelligible to their contemporaries' (E. F. Sutcliffe, *A Catholic Commentary on Holy Scripture*, 1953). This, as will be seen, later allowed a much broader view of Genesis than might originally have been thought possible. It should also be kept in mind that these pronouncements of the churches were primarily directed, not against Darwinism, but against the historical and textual criticisms of the Bible, which were giving rise to some wild and heretical fancies that later research has shown to be unwarranted.

Only the Darwinian challenge to the first three chapters of Genesis concerns us here. By some, including as already mentioned, W. E. Gladstone (in the *Nineteenth Century*, 1886), the attempt was made to reconcile the scientific and biblical accounts of creation. 'The five origins, or first appearances of plants, fishes, birds, mammals and man,' wrote Gladstone, 'are given to us in Genesis in the order of succession, in which they are also given by the latest geological authorities.' He added that 'these astonishing anticipations were a God-given supply' and 'entirely transcended, in kind even more than in degree, all known exercise of human faculties.' Such a view, which I have heard affirmed even in 1955, cannot be sustained. Thus according to the first chapter of Genesis, fish and birds were created on the same day, mammals a day later, whereas biologists would have put the birds close in time to the mammals and far from the fish. Further, grass and fruit trees were created a day before the sun. Actually, Gladstone's view had been rejected fifty years earlier, by Dean Buckland in his *Bridgewater Treatise* of 1836, already cited. Moreover, even Gladstone's view is opposed to a literal reading of Genesis since it allows the 'day' an unusual length.

Darwinism also controverted the fixity of species, which Genesis was supposed to imply, though in fact it is vague on the point. In earlier times Christians, like everyone else, thought that various animals arose from dirt, or changed from one kind into another. The fixity of species was chiefly a postulate of the seventeenth and eighteenth centuries, and owes much to the formal classification of animals by Ray and Linnaeus; though, contrary to what is usually stated, Linnaeus himself gave up this belief later in his life (J. Ramsbottom, *Proc. Linn. Soc. Lond.*, 150 : 192–219, 1938). It has even been suggested, how rightly is hard to say, that the idea that each species was specially and individually created by God became linked with Christian belief through the popularity of Milton's *Paradise Lost* (A. Moore, *Science and Faith*, 1889).

While Darwinism was widely supposed to contradict the accuracy of the Bible, what it actually challenges is the literal rendering of the first three chapters of Genesis, and if these are properly to be regarded as allegorical, no conflict need arise. These chapters were, however, considered to be literally true by nearly all Christians in the first half of the nineteenth century, and this despite the fact that the two accounts of the creation, in the first and second chapters respectively, are very different and in part contradict each other. For instance, birds arose from the waters in Chapter 1 (v. 20) and from the ground in Chapter 2 (v. 19). Actually some of the early fathers of the Church, and various later Christians, treated at least part of the account as allegorical. St. Augustine, for instance, held that the earth, with all in it, was created in one instant, and that the six days were introduced to help the mind to grasp this concept (quoted by J. C. Hedley, *Evolution and Faith*, *repr.* 1931). Irenaeus, Clement and Athanasius likewise referred to the Fall in allegorical terms (C. Gore, *Lux Mundi*, 12th ed., 1891, app. 2). At the end of the fifteenth century, Dean Colet, one of the early reformers in England, regarded Genesis as a poetic fiction conveying essential truths (H. G. Wood, *Belief and Unbelief since 1850*, 1955). Sir Thomas Browne considered that 'unspeakable mysteries in the Scriptures are often delivered in a vulgar and illustrative way; and being written unto man, are delivered, not as they truly are, but as they may be understood.' (*Religio Medici*, 1643).

A solution along such lines is now agreed among all Christians except the few extreme fundamentalists. In this matter Roman Catholics have tended to be more conservative than Anglicans, but Pius XII recently pronounced in the encyclical *Humani generis* (trs. R. A. Knox, 1952) that 'the Teaching of the Church leaves the doctrine of Evolution an open question, as long as it confines its speculations to the development, from other living matter already in existence, of the human body. (That souls are immediately created by God is a view which the Catholics faith imposes on us.)' And as regards the early chapters of Genesis, 'although it is not right to judge them by modern standards of historical composition, such as would be applied to the great classical authors, or to the learned of our own day, [they] do nevertheless come under the heading of history. . . . These chapters have a naif, symbolical way of talking, well suited to the understanding of a primitive people. But they do disclose to us certain important truths.'

The true significance of the first chapter of Genesis is to assert that God made the universe and all in it, that He saw that it was good, and that He placed man in a special relationship to Himself. These three assertions seem far in advance of any pagan myth, and may well be claimed as 'astonishing anticipations', to use the phrase which Gladstone unjustifiably applied to another part of the narrative.

(Gladstone was on sounder ground in accusing T. H. Huxley of thinking Genesis to be a lecture, whereas he thought it a sermon.) The three basic assertions do not conflict with Darwinism. Evolution is concerned with secondary derivations and not with the creation of something out of nothing; the assessment of the world as good is not one with which science is concerned; and man's spiritual relationship with God seems little affected by whether he evolved from beasts, as Darwinists claimed, or was made from dust, as stated in Genesis (*cf.* A. Moore, *Science and Faith*, 1889).

The first chapter of Genesis was written in the fifth century B.C., while nearly all of the second and third chapters are thought to have been set down in about the seventh century B.C., through the combination of two yet earlier narratives. There is no reason for thinking that the Hebrew writers of these periods would be concerned to give scientific or historical facts as we understand these terms today, and every reason to suppose that 'the sacred author adopted a style of writing, recognizable as such by his contemporaries, by which he clothed important religious truths in the form of a concrete graphic narrative' (E. F. Sutcliffe, *A Catholic Commentary on Holy Scripture*, 1953). It may further be claimed that the spiritual truths concerned could not have been expressed or apprehended except in such terms. On this interpretation, Christians can accept the theory of animal evolution without compromising their belief in the accuracy of the Bible.

While the historic conflict between Darwinism and Genesis was centred upon the first chapter, the difficulties presented by the third chapter are at least as great. The snake and the fruit-tree (an apple by tradition, but not in Genesis) can be treated as imagery rather than facts without any loss in the spiritual meaning of the story, which is that man was intended to be without sin, but disobeyed God's will. The post-Freudians, like certain early Christian writers, have supposed the Fall to be concerned with man's acquisition of sexual knowledge.[1] But this can hardly have been meant by the biblical writers, since already in the first chapter Adam and Eve were bidden to be fruitful and multiply, and in the second chapter were described as one flesh. The story concerns problems of human conduct more fundamental than those relating to sex, including why, knowing the good, man yet follows the evil.

Many Protestants now regard the first three chapters of Genesis as poetic imagery, but Roman Catholics tend to treat them as allegorical history (*Humani generis* already cited; H. J. T. Johnson, *The Bible and the early History of Mankind*, 1947; E. F. Sutcliffe, *A Catholic Commentary on Holy Scripture*, 1953). On the Catholic view, Genesis describes events that really took place, though not in the form in which they are pictured. The reasonableness of this view depends on how much is to be treated as fact and how much as imagery. Citing from the biblical commission of 1909, Father Johnson stated that the following six points in the story of the Fall are to be interpreted strictly: the unity of the human race; the original happiness, integrity and immortality of our first parents; a precept given by God to man to prove his obedience; its transgression through

[1] Man's deep-rooted fear of snakes has sometimes been attributed to a snake symbolizing the male sex organ. This fear is already present in small children (H. Prachtl, *Wien Zeits. Phil. Psych. Päd*, 2:68–70, 1950); it is also found in birds (A. L. Rand, *Bull. Amer. Mus. Nat., Hist.* 78:232–5, 1941), yet birds do not possess a penis. In both birds and men there would be survival value in evolving an innate fear of a dangerous enemy. Hence while man's fear of snakes may occasionally be neurotic, in general it seems healthy and beneficial.

persuasion by the devil; a resulting fall from primeval innocence; the promise of a future Redeemer. Only one of these points, the postulated unity of the human race, is strictly biological, so the others will be deferred to later chapters. The unity of the human race has been reasserted for Roman Catholics in *Humani generis*: 'Christians cannot lend their support to a theory which involves the existence, after Adam's time, of some earthly race of man, truly so called, who were not descended ultimately from him, or else supposes that Adam was the name given to some group of primordial ancestors.'

As stated in the last chapter, new species of animals normally arise from isolated populations, not individuals. But it is theoretically possible for one pair to give rise to a new species, and this may well have happened in various land animals that have found their way to remote islands. Hence on biological grounds it is not at all impossible, though it would be unusual, if the population ancestral to man were at one time reduced to a single pair through mortality or, more probably, emigration. A greater difficulty is that the structural change from animal to man must have involved a large number of generations, so that at the biological level no one generation could be named Adam and Eve in the sense that before them all were animal and after them all were human. But on the spiritual plane there could be such an absolute difference, if man is primarily to be distinguished from beasts by his possession of a soul, as a single act of creation by God: this is a matter on which science has nothing to say. Neanderthal man was contemporary with *Homo sapiens*, so that if the view put forward in *Humani generis* is followed, Adam presumably lived before the divergence between *Homo neanderthalensis* and *Homo sapiens* took place. But it is agreed that these two forms had a common ancestor, and although they are usually treated as separate biological species, they possibly interbred, and were possibly races of one species, as mentioned in the last chapter.

To conclude, the available biological and fossil evidence would allow the unity of the human race as laid down in *Humani generis*. At the same time, the evidence is so scanty that nearly all biologists, and I imagine most Protestants, would prefer to leave the question open. In any case, it seems not unreasonable to suppose that when the biblical writer used the names Adam and Eve for the first of mankind, he was unconcerned as to whether they really existed as a pair of historical persons. The reason that this latter view is maintained in *Humani generis* is theological, not scientific or historical, being connected with Original Sin, as will be discussed later.

The fear that led to the Oxford Declaration cited earlier was that, if any part of the Bible were shown to be inaccurate, all might be doubted; and the distrust of Darwinism was deepened because its leading exponents, T. H. Huxley in Britain and Ernst Haeckel in Germany, went on to attack Christian belief on other grounds. Huxley, in particular, wrote strongly against the miracle stories in the Bible. But while science can refute the literal truth of Genesis, it cannot refute miracles, since these, by definition, are outside the laws of nature. It is legitimate to argue whether anything can happen, or can be said to happen, contrary to natural laws, but the reasons for or against this view are not scientific. It is also true that the credulous of all faiths have been too prone to seek for supernatural causes in strange events, and that in earlier times, when the facts of science were less well known, happenings were accepted as miraculous which we, with fuller knowledge, might have been able to explain through natural causes. These trends are admitted by the Church, which has tried, though perhaps not with full success, to restrict miraculous interpretation to rare events that were truly contrary to the laws of nature. The great miracles of the Resurrection and the Virgin Birth are not made significantly less likely by our fuller knowledge of biology, since the facts of conception and death were already known well enough for this purpose two thousand years ago. The evidence for or against them as miracles depends on history and also, if that claim be allowed, on spiritual insight. My concern here is not to discuss how miracles, or alleged miracles, should be interpreted, but to point out that the basic issue is outside science, and hence outside the scope of this book. It may be added that the published opinions of T. H. Huxley and Ernst Haeckel suggest that, even if Darwin's discoveries had not been made until after their deaths, they would still have attacked miracles, so that the linking of such views with Darwinism was accidental.

As already mentioned, the conflict between Darwinism and the truth of Genesis coincided with a much more serious challenge to the trustworthiness of the Bible on historical and textual grounds (see e.g. L. E. Elliott-Binns, *Religion in the Victorian Era*, 1936). This problem is also outside the terms of reference of this book, so cannot be followed further, though a brief comment may be added. As I understand the position, it is now agreed that the earlier critics went much further than was justified. But because orthodox theologians carefully examined the new evidence and ideas, the historical and textual accuracy of the Bible now rests on a much surer, because more critical, basis than before, and researches that were once dreaded as hostile are now welcomed as helpful to the Christian faith. The idea that every word of the Bible is literally true is not, of course, maintained.

4

Natural Selection

The concept of natural selection was the most original part of Darwin's great book and gave it needed strength, since for the first time an effective means was put forward whereby evolution could have taken place. In each species of animal and plant, far more new individuals are produced than can survive to breed. Hence there is a huge annual mortality, and while much of this falls at random, there must be a tendency for more of the better adapted and fewer of the less well adapted individuals to survive. As a result, those individuals that are better adapted tend to leave proportionately more offspring than the others, and any hereditary advantages that they possess tend to be passed on to the next generation and to spread through the population.

In Darwin's time, little was known about heredity, and the far-reaching discoveries of the Abbot Mendel at Brünn were overlooked until the start of the twentieth century. Mendel showed the way in which mutations (hereditary changes) present in plants are passed on to their progeny, but the idea that such mutations could form the basis of natural evolution was for a long time but dimly realized, for two main reasons. First, mutations producing differences from the normal or wild type of an animal or plant are nearly always harmful in nature. Secondly, they are nearly always recessive, meaning that they have a visible effect only in those offspring which receive them from both parents. If received from only one parent, their effects are suppressed by the dominant wild type.

Thirty to forty years ago, there was also a strong reaction against the concept of natural selection, due to a number of causes, not least of which was an over-enthusiastic advocacy of it by the previous generation of naturalists. This led to uncritical and sometimes absurd instances of alleged adaptation. It was claimed, for instance, that Flamingoes were pink so as to conceal them against the sunset. There was also no direct positive evidence for natural selection, while it was hard to understand how complex interlocking adaptations could have been evolved by its means; and, at the same time, closely related species were often separated by slight and apparently non-adaptive differences.

The last thirty years has seen a reversal of this opinion, and the vindication of the theory of natural selection, due to many research workers, and especially to R. A. Fisher, whose book *The Genetical Theory of Natural Selection* (1930) began the new period. The main advances came, as so often in science, not through the study of grand and intricate adaptations, but by the measurement of small and seemingly trivial differences. One of the most important ideas, now fully established, is that each hereditary element, or gene, has not merely one but many effects, and each part of the body is affected not merely by one but by many genes. It is therefore possible for any harmful effects of one gene to be reduced or neutralized by other genes, and any beneficial effects to be strengthened or increased. Those types of individual tend to leave most offspring which carry the most favourable combinations of genes for survival, and natural selection works on the whole complex of genes.

Mutations, the occasional changes in the genes, are usually harmful and usually recessive, as already mentioned. But these facts, once so puzzling, can now be explained through natural selection. In general, the normal or wild type of an animal has been subject to stringent natural selection in the past, so that any mutational change away from this type is likely to be harmful in nature. When this is so, natural selection will favour those modifying genes which reduce or neutralize the harmful effects of the mutation in question, and often they reduce them so much that they are suppressed in an individual receiving the mutation from only one of its two parents. This is what is meant when the mutation is described as recessive to the normal type. That such recessiveness is due to selection has been proved in the laboratory, where it has been possible by breeding for several generations to change the effect of a mutation from dominant to recessive, or conversely. Speaking more fully, it has been possible through selection of the breeding stocks to provide different combinations of genes modifying the character concerned. An important conclusion of this research is that, because each character is affected by many genes, natural selection can produce evolutionary changes much more rapidly and surely than was at one time thought possible.

In recent years, several evolutionary changes due to the natural selection of a mutation have been demonstrated in the wild. Just over a century ago, an unusually dark (melanic) form of the Peppered Moth was recorded near Manchester, and it has since spread rapidly in industrial areas, though not in rural areas. Through breeding experiments, it is known that the melanic form depends on a hereditary mutation, and on tree-trunks darkened by soot it is much less conspicuous to the human eye than the normal pale form. Moreover, when many individuals of both the melanic and pale forms were simultaneously released in an urban wood, wild birds took many more of the pale than the melanic form, whereas the reverse happened when both forms were released in a rural wood. It may therefore be concluded that the evolution and spread of the melanic form in industrial areas has been due to natural selection, in this instance caused by birds (H. B. D. Kettlewell, *Proc. Roy. Soc.* B,145 : 297–303, 1956).

Again, in Californian citrus groves a mutant of a scale-insect has been evolved which is highly resistant to the cyanide fumigation to which these orchards are subjected. There are also some beautiful examples in bacteria in which a mutant has spread because it is much more resistant than the normal type to a poison or bacteriophage (T. Dobzhansky, *Genetics and the Origin of Species*, 3rd ed., 1951). The last-named author also reported that on a Californian mountain there occur two forms of a species of fruit-fly (*Drosophila*), one of which decreases from March to June and then increases till October, while the other increases from March to June, and then decreases till October. The differences between the two forms are hereditary. The release and later recapture of marked individuals of both types in the wild showed that the changes in their numbers at different seasons were caused by differences in their mortality-rates. Further, when both forms were reared together in competition with each other in cages, one came to predominate at lower and the other at higher temperatures. It may therefore be concluded that the seasonal changes in their abundance in the wild are a result of natural selection, one being the more successful in colder weather, and the other in warmer weather. In melanic moths, the critical adaptation is one of colour, and so is obvious to the human eye, but in this fruit-fly there was no conspicuous difference between the two forms. This shows how dangerous it can be to claim that the differences between two forms are non-adaptive, as was frequently done by critics of natural selection some thirty years ago.

These instances show that, with a slight change in the external conditions, such as darker tree-trunks or warmer weather, natural selection may cause the spread of hereditary adaptive changes through a population. On the other hand, where the conditions to which an animal is adapted do not change significantly, natural selection is equally important as a conservative force, keeping each form adapted to its normal environment by eliminating the unfavourable mutations that repeatedly arise. It is for this reason that each form remains so constant in appearance. Nevertheless, the capacity for change is there. The change from pale to dark colouring in melanic moths is slight when compared with the big changes in many animals revealed by the fossil record, but it has taken place in less than a century, a period which on the geological time-scale is negligible. The influence of natural selection in producing large changes has not been proved, but there has not been time to see it. As Darwin pointed out, the speed at which big hereditary changes can take place is well shown by man's selection of his domestic and cultivated stocks. Combining all the available evidence, nearly all biologists are now agreed that natural selection is capable of producing the marked evolutionary changes that have occurred in the long course of geological time.

It has sometimes been doubted whether natural selection could be sufficiently 'creative', but in this connection a parellel by G. G. Simpson (*Scientific Monthly*, 64 : 481–95, 1947) may be helpful. Suppose, he said, that a huge bag contained all the letters of the alphabet in equal abundance, and you drew out at random three letters at a time, with the aim of getting the word 'cat', discarding the letters if they did not form this combination. You might spend days, weeks or years before you achieved your aim, indeed you might draw all the letters from the bag without achieving it. But suppose that, each time you drew out 'c' or 'a' or 't' you could replace it in the bag, discarding other letters as before, then your chances of getting the right combination would be greatly increased. They would be further increased if, when you drew out two of the right letters at one time, you could clip them together before returning them to the bag. 'Cat' is a highly improbable combination to appear at random, but by selecting in the way described you would be certain to draw it from the bag, and since the word 'cat' did not exist before, this type of selection can be called creative.

Simpson stressed that his analogy was over-simplified, but provided that it is not pressed too far, it is valid biologically. 'Creative natural selection works in a similar but vastly more complicated way', since the number of possible combinations of genes, even in one simple organism, is enormous, and much of the mortality in each generation falls at random. For a deeper analysis of the problem, the reader may be referred to R. A. Fisher's short *Creative Aspects of Natural Law* (1950). Simpson (*The Meaning of Evolution*,

1951) has further stressed that the 'struggle' for existence is a somewhat misleading term, since the critical factor is the number of offspring that survive, and there is not usually a struggle or battle in the ordinary sense of the term. To conclude, the theory of natural selection, though much modified and extended, remains in essence as Darwin formulated it, and biologists are in general agreed that natural selection has been the main agent of evolutionary change. Further evidence for this view will be given in the next chapter.

5

Creative Evolution

The first phase of the attack on Darwinism concerned evolution as a fact, and quickly resulted in victory for the Darwinists. The second phase, a more serious attack, was concerned with the means of evolution. Mutations occur, or seem to occur, at random, and natural selection likewise, so that many felt that evolution by natural selection was insufficient to explain the intricate adaptations and purposiveness of living things. A doubt on these grounds was expressed only two years after the publication of *The Origin of Species*, by Sir John Herschel, who wrote that he could not 'accept the principle of arbitrary and casual variation and natural selection as a sufficient account, *per se*, of the past and present organic world', and that 'an intelligence guided by a purpose must be continually in action to bias the directions of the steps of the change' (*Physical Geography of the Globe*, 1861 ed., cited in *More Letters of Charles Darwin*, 1903).

On Lamarck's view, put forward before Darwin's time and accepted as a partial explanation by Darwin himself, 'purposive guidance' comes from the animal, and adaptive evolution occurs through the use and disuse of organs during the animal's lifetime, such tendencies being passed on to its offspring. But although there have been many later advocates of this view, extensive research has yielded no positive evidence whatever that characters acquired during an animal's lifetime can be inherited, and the indications are now overwhelmingly strong that they are not so inherited. The views of the Neo-Lamarckists have been summarized by P. G. Fothergill (*Historical Aspects of Organic Evolution*, 1952) and a last flicker came in 1953 (F. Wood Jones, *Trends of Life*; see also *Nature*, 173 : 51). At the present day, the concept of the inheritance of acquired characters has been generally discredited and rejected.[1]

While the Lamarckists supposed that adaptations could be evolved through a striving or urge of the animal itself, later critics of natural selection postulated instead that a force outside the animal provides the driving impulse and purposive direction of evolution—a Universal Mind, a Life Force, Creative Evolution, Emergent Evolution, Holistic Urge—various terms have been used. The force in question has by some writers been conceived as natural and contained within the structure of a godless universe, and by others as mystical or supernatural, though not necessarily of divine origin. Such views have been advanced by both atheists and Christians, some of whom have used philosophical rather than biological arguments. The philosophical problems will be examined later, and the present chapter is concerned solely with the biological evidence.

Apparently the first biologist to advocate such a view, though in vague terms, was Mivart, who held that natural

[1] Some of the apparent instances of Lamarckian inheritance were perhaps due to the selection of hereditary factors which replaced or reinforced direct phenotypic (non-hereditary) variations, thus simulating Lamarckism (G. G. Simpson, The Baldwin effect. *Evolution* 110–17, 1953).

selection was inadequate, and that 'an internal power or tendency is an important if not the main agent in evoking the manifestations of new species on the scene of realized existence' (*The Genesis of Species*, 1871). The idea was then developed and popularized by Samuel Butler, who had a remarkable career. After being dismissed from home by his father, a parson, he became a successful sheep-farmer in New Zealand, then returned to England, became a painter, also an amateur musician and composer, and wrote two famous works of fiction, his utopian *Erewhon* and an autobiographical novel, *The Way of All Flesh*. He gave up his career as a painter and considered so strongly that both Darwinism and the Church should be opposed that he spent much of his later life writing vigorously against them. His basic criticism of natural selection is implied in the title of one of his books, *Luck or Cunning as the Main Means of Organic Modification?* (1886), luck standing for natural selection and cunning for some form of evolutionary urge. 'Shall we maintain,' he wrote, 'that the eagle's eye was formed little by little by a series of accidental variations, each one of which was thrown for, as it were, with dice? We shall most of us feel that there must have been a little cheating somewhere with these accidental variations before the eagle could have become so great a winner.' (*Evolution Old and New*, 1879.)

Samuel Butler's views were ignored by biologists. They were also ignored by most other people, until powerfully advocated by Bernard Shaw, notably in his Preface to *Back to Methusaleh* (1921). Shaw, like many others, was chiefly concerned about the moral and philosophical implications of natural selection, discussed later. On the scientific side, criticisms essentially similar to those of Samuel Butler have been voiced up to the present day. Thus Canon C. E. Raven, in his Gifford Lectures, stated: 'It has always been enormously difficult to maintain, for example, that the primitive bird Archaeopteryx had been evolved at random and by the requisite astronomical number of variations, each by itself a hindrance, which gradually transformed arms into wings, provided a sternal keel and girdle, the aeration of bones and skin-sacs, the warm blood, the feathers, and all the essential modifications of reptilian structure.' The same writer argued that 'a sequence of at least five distinct events, outside the run of normal behaviour, and structure' had to take place before the Cuckoo could successfully leave its eggs in the nests of foster-species for them to rear. 'Yet each by itself is useless. . . . The odds against the random occurrence of such a series of coincidences are astronomical' (*Natural Religion and Christian Theology*, 1953). Again, 'complex structures and habits, by their very nature, cannot have been built up step by step' (*Science, Religion and the Future*, 1943).

To the same effect, Canon Smethurst (*Modern Science and Christian Beliefs*, 1955) wrote 'it is exceedingly doubtful whether pure chance could ever have produced such co-ordinated or beneficial developments as occur in a living organism.' Likewise, the agnostic Jacquetta Hawkes (*Man on Earth*, 1954) thought that even in millions of years sexual selection by the hen bird could not conjure up so wonderful a creation as the male Argus Pheasant. Again, E. L. Grant-Watson, writing of the Large Blue Butterfly whose larvae are housed by ants, found it impossible 'to imagine that the many synchronizing adaptations are the result of small continuous variations, or, for that matter, of larger mutations.' (*Guild of Pastoral Psychology*, Lecture

82, 1954.) Similar views were advocated by J. C. Smuts (*Holism and Evolution*, 1926). They have been far more frequent among general writers than zoologists, but the professor of zoology at Cambridge recently remarked that 'no amount of argument, or clever epigram, can disguise the inherent improbability of orthodox [Darwinian] theory. . . . There will always be a few [biologists] who feel in their bones a sneaking sympathy with Samual Butler's scepticism' (J. Gray, *Nature*, 173 : 227, 1954).

I have set out these criticisms at length because, although most biologists have dismissed or ignored them, they are popular among those who are not biologists, particularly among those who, like Shaw and the followers of Jung, have rejected belief in the Christian God but feel that there must be a purpose in the universe; and in recent years several Anglican theologians have also supported them.

These criticisms may be summarized as saying first that adaptations are too complex and interlocking to have been brought about by so random an agent as natural selection, and secondly that necessary intermediate steps in the gradual evolution of such adaptations could not be advantageous. What, then, is put in the place of natural selection? H. Bergson (*Creative Evolution*, trs. A. Mitchell, 1911) wrote: 'If the accidental variations that bring about evolution are insensible variations, some good genius must be appealed to—the genius of the future species—in order to preserve and accumulate these variations, for selection will not look after this. If, on the other hand, the accidental variations are sudden, then . . . all the changes that have happened together must be complementary. So we fall back on the good genius again, this time to obtain the convergence of simultaneous changes.' Bergson called the good genius the 'élan vital', and Bernard Shaw and other writers have likewise advocated a Life Force under various names. Similarly Canon Smethurst wrote 'Where there is progress, order, and direction, the evidence favours the view that there is a guiding, rational mind at work behind it', and Canon Raven stated that natural selection 'sifts and fixes and educates, and in small ways develops; there is no evidence at all that it inaugurates or creates. . . . Progress, the emergence of novelty, manifests an urge towards fuller and more complex achievement and (it seems evident) some co-ordinating "Organizer" or holistic principle which enables simultaneity and harmonious change' (*Science, Religion and the Future*, 1943). More crudely, I have heard it said that since natural selection acts merely as a sieve, the direction of evolution is determined by the mutations, which in turn are regulated by God.

Such views have a popular appeal, so it may be worth saying at the start that of the many biologists studying evolutionary processes at first hand at the present day, none (so far as I am aware) gives credence to a Life Force or similar agent, and that this holds for both the agnostics and the Christians among their number. As such views were advanced more than eighty years ago, there has been ample time for biologists to appraise them.

The Life Force was postulated owing to the seeming inadequacy of natural selection, but it does not in itself help in the analysis of the means of evolution. It measures nothing, and observations and experiments cannot be designed to test its truth or falsehood. Indeed, some of its adherents have claimed that it is outside the field of scientific inquiry. It would appear to be just a name for processes that are not understood, or more accurately, that were not understood, and it may be compared with the old gods of fire, or wind, or lightning, which were created by primitive men to account for mysterious natural forces.

Moreover a basic error is involved in this concept, since the Life Force is held to set the course of evolution through its influence on the direction of the mutations. If natural selection were 'merely a sieve', creative change would presumably be due to the direction of the mutations. But in fact, the mutations are random in relation to the needs of the animal, many more of them being harmful than helpful, and in addition they occur much too infrequently to be by themselves the cause of evolutionary change. Whatever be the agent responsible for the directions of evolution, it must act after the mutations have occurred, that is to say there must be selection from among the mutations, and the only form of selection known to occur is what Darwin termed natural selection. In illustration of this point, R. A. Fisher pointed out that polydactyly, the production of an extra toe, is a frequent mutation in vertebrate animals, including man, yet in no instance has it led to the evolution of a six-toed animal. The determining factor in this evolution has not been the mutation, but its subsequent elimination by selection (*Creative Aspects of Natural Law*, 1950, also in *Evolution as a Process*, ed. J. Huxley, *et al.*, pp. 84–98, 1954; two illuminating papers in relation to this and the previous chapter).

It will be noticed that the critics of natural selection have chosen intricate adaptations for discussion, whereas in general we should seek to understand the simple before tackling the complex. As discussed in the last chapter, the modern knowledge of genetics and natural selection suffices to explain simple steps in evolution. Complex steps, such as the evolution of an eye, cannot be directly observed, as they must have taken place over millions of years. Nevertheless, Darwin himself anticipated his critics on this example, pointing out that there are in nature numerous gradations from a simple to a complex type of eye, and also that the eye is subject to inherited variations, so that 'the difficulty of believing that a perfect and complex eye could be formed by natural selection, though insuperable by our imagination, should not be considered as subversive of the theory.' In quoting this, R. A. Fisher (in *Evolution as a Process*) stressed that the difficulty in such cases is one of the imagination and not of the reason.

Take again the case of Archaeopteryx, the fossil bird-reptile from the Jurassic discussed by Canon Raven. This was recently described as 'probably the most precious, the most beautiful, and the most interesting fossil hitherto discovered' (G. R. De Beer, *Adv. Sci.*, 42, 1954, from whom the following account is largely taken). Archaeopteryx was transitional between a reptile and a bird, but like other 'missing links' it was not intermediate in every character but was a mosaic, with some primitive and some specialized features. Thus it was wholly reptilian in its long tail of twenty free and unfused vertebrae, in the method by which the vertebrae of the spine articulated, in the free unfused metatarsal bones of the foot and metacarpal bones of the forelimb, in the fingers bearing claws, also in its sacrum, ribs, simple brain with small cerebellum, and the absence of a keel to the sternum. It also differed from all living birds in having teeth. On the other hand, it was typically avian in having feathers, in having the two collar-bones fused to form a furcula or wishbone, in having the pubic bones of the pelvis pointing backwards not forwards, and in having the big toe pointing in the opposite direction to the other three. Further, the feathers were of the same structure, with rachis and barbs, as in modern birds, and were arranged in the same manner to form a wing.

These points show that Canon Raven chose an unfortunate example for his thesis. In particular, Raven regarded the evolution of a sternal keel as essential to Archaeopteryx. When he wrote, the sternum of Archaeopteryx had not been found, but new photographs of the fossil, using ultra-violet light, have now revealed that it did not have a keel. This means that it lacked strong breast muscles and so must have glided, not flapped its wings. This is confirmed by the small size of the cerebellum, the part of the brain which controls balance and the complicated movements of a flying bird. In view of the evidence, it is reasonable to claim that flying birds could have evolved from flightless reptiles in a series of steps. Moreover some of these steps were fully evolved before others began, so that there is no need to invoke the intricate synchronization of adaptations that Raven supposed was needed.

Similar criticisms can be made of the other examples selected by the advocates of Creative Evolution. If we had only the Argus Pheasant, it would be hard to conceive how sexual selection could have led to such elaborate and beautiful colour patterns, with such complex behaviour to display them. But among birds there is every gradation, from species in which the males are as dull in colour as the females, to those with small bright areas and simple displays, to those with more elaborate structures and behaviour, leading up to the glories of Peacock and Argus Pheasant. Again, the European Cuckoo selected by Canon Raven is the most specialized of all cuckoos. There are other species, as successful in their own way, which have only one or two of the features which Raven regarded as essential. Thus various species do not have abnormally small eggs, but lay in the nests of birds of their own size, in many species the young cuckoo does not eject the young of its host but is raised with them, and in some forms the egg bears little or no resemblance in colour to the eggs of the host. Hence there is little difficulty in seeing how the parasitism of the European Cuckoo could have been evolved in a series of gradual and functional stages.

Mimicry has also been found difficult. The perfect resemblance of certain harmless butterflies to others that are poisonous, or of the cuckoo's egg to that of its foster-parent, would, it is argued, be valueless if incomplete, and yet could not be evolved in one large mutation. But mimicry has been evolved to counter the discrimination of bird enemies, and if one supposes that the bird's power of discrimination (often innate) was evolved gradually, no difficulty need arise. Animals do not normally evolve innate responses unless they have survival value. Thus a gull will unconcernedly brood a rubber ball, a square brick or a watch placed in its nest; for in its normal life, failure to distinguish such objects from its own eggs does not result in any appreciable disadvantage, so that it has not evolved special behaviour to cope with them. It may therefore be surmised that at one time songbirds were likewise unable to detect a strange egg among their own; but if the presence of a strange egg led to the starvation of their young, as it would with a cuckoo's egg, there would be survival value in their detecting it, and an innate power of discrimination might then be evolved. But at first the cuckoo's egg would be so unlike their own that only a crude discrimination would be needed. If this were enough for detection of the strange egg, and were innate, natural selection would favour its evolution. At this stage, but not before, there would be survival value to the cuckoo in a mutation increasing the resemblance of its egg to that of its host. When this was enough to deceive the host, there would be survival value to the host in improving its discriminatory power, and so on, until both accurate discrimination and a strong mimetic resemblance had been evolved.

During the nineteenth century, in opposition to the over-simplified mechanistic theory of living matter, vitalists stressed the purposiveness in structure and behaviour of animals and plants (for a recent exposition, see E. S. Russell, *The Directiveness of Organic Activity*, 1945). This seeming purposiveness has sometimes been adduced as a further argument against natural selection and in favour of a guiding rational or spiritual influence in evolution (A. F. Smethurst, *Modern Science and Christian Beliefs*, 1955). But the quality of 'purposiveness' or 'directiveness' has not been precisely defined, and seems incapable of analysis, and it may well mean no more than extreme intricacy, combined with structures and behaviour closely adapted for survival by natural selection. Russell and Smethurst have claimed that this quality of living matter lies outside science, but scientists may reject the intrusion of pseudo-mystical claims into their proper field of observation. Such rejection does not at all imply, as sometimes supposed, that scientists fail to recognize the marvellous organization of the living plant and animal, or that they think this organization has been adequately explained in the known terms of physics and chemistry.

Smethurst added that the rate at which complex adaptations

have been evolved has been too slow for natural selection, and that this likewise requires the postulate of a creative Mind. But, that the difference between natural selection and spiritual control is one of speed, may be doubted not only on scientific but also on theological grounds. While the matter cannot be proven, nearly all biologists consider that the available evidence favours the view that natural selection has had time enough to produce the major changes of evolution. The difficulty of Smethurst and others is partly due to their considering natural selection a matter of 'pure chance', a point discussed in the next chapter.

Summing up, the evidence suggests that the evolution of complex adaptations could have been gradual, since despite what has been claimed, the intermediate steps could have been functional. Further, interlocking adaptations need not have been evolved synchronously, since 'missing links' are often a mixture of specialized and primitive features. Moreover, mutations occur at random in relation to the needs of the animal, so that the directions of evolution are determined, not by a force governing the mutations, but by their subsequent selection or rejection, for which purpose natural selection seems adequate. Hence the concept of a Life Force or holistic urge, which at best was a term naming what could not be explained, may be rejected as unnecessary and misleading.

6

Chance or Plan?

It is not self-evident that God exists, and recognition of this has prompted various attempts to prove God's existence by rational argument. In the first half of the nineteenth century, the one widely accepted argument of this nature was that popularized by William Paley in his *Natural Theology* (1802), that animals and men are so wonderfully made that they must have had a Designer. The mechanistic laws on which the physical universe was then conceived as being run were deemed quite inadequate to account for the intricate purposive adaptations of animals, and these, it was argued, must have been specially created for the purposes which they serve.

In a vivid style, Paley piled up instance after instance of complex and interlocking adaptation in the body. Having described, for instance, the structure of the gullet and windpipe, he added: 'Reflect how frequently we swallow, how constantly we breathe. In a city-feast, for example, what deglutition, what anhelation! yet does this little cartilage, the epiglottis, so effectually interpose its office, so securely guard the entrance of the windpipe, that whilst morsel after morsel, draught after draught are coursing one another over it . . . (which, nevertheless, must be opened for the breath every second of time) . . . not two guests are choken in a century.'

One result of Darwin's theory was to show that the adaptations of animals could have originated by natural means, and that there was no need to invoke a special creation for them, but Paley's argument from design had been so widely accepted that this caused a great shock at the time. Actually, Paley's views can be questioned not only on biological but also on theological grounds, as by F. Temple, one of the first Anglican churchmen to accept the theory of evolution, and later Archbishop of Canterbury: 'It seems in itself something more majestic, something more befitting Him to Whom a thousand years are as one day and one day as a thousand years, thus to impress His Will once for all on His creation, and provide for all its countless variety by this one original impress, than by special acts of creation to be perpetually modifying what He had previously made' (*The Relations between Religion and Science*, 1884). In view of the arguments considered later in this chapter, another quotation from Temple may be added, included by him in a sermon at Oxford during the meeting of the British Association in 1860, on the day after Huxley and Wilberforce held their famous debate. 'It has been common to trace the power of God not in that which is universal, but in that which is individual: not in the laws of nature, but in any apparent interference with those laws' (*The present Relations of Science and Religion*, 1860; Temple was, of course, arguing against such views).

Some Christians, less wise than Temple, have tried to retain the essence of Paley's argument by regarding natural means as sufficient to account for part, but not the whole, of the evolutionary process. Thus, as mentioned in the last chapter, it has been claimed that since the cause of mutations is unknown, their direction is determined by

God; and the concept of the Life Force has sometimes been upheld as a more subtle expression of the same idea. But as shown in the last chapter, mutations do not in themselves determine the direction of evolution, and even if they did, such a method of arguing would be wrong. So often in the past, the appeal to a divine or mystical agency has merely covered ignorance of scientific facts, and because Christians (though not only Christians) have argued in this way, they have earned deserved disrespect from scientists. Usually, the unknown factor for which a supernatural agent was invoked has later been shown to have a natural cause.

Other Christians have fully recognized the danger. As Henry Drummond wrote: 'There are reverent minds who ceaselessly scan the fields of nature and the books of science in search of gaps—gaps which they fill up with God. As if God lived in gaps' (*The Ascent of Man*, 1894). Likewise Charles Kingsley said in a letter of 1863 welcoming Darwin's book that we now 'have to choose between the absolute empire of accident and a living, immanent, ever-working God '(*Life*, 1877). To cite a modern scientist, who is also a Christian: 'When we come to the scientifically unknown, our correct policy is not to rejoice because we have found God: it is to become better scientists, and to think a bit more deeply until we can devise some model, or some concept, that will bring the previous unknown into the pattern of the known' (C. A. Coulson, *Science and Religion, a Changing Relationship*, 1955).

On the views so far considered, evidence for purposiveness was sought in animal adaptations. Alternatively, the argument has been transferred from the nature of the organic to the nature of the inorganic world, notably by L. J. Henderson (*The Fitness of the Environment*, 1926; also by C. F. A. Pantin, *Adv. Sci.*, 8 : 138–50, 1951). Thus it has been claimed that life would not have been possible without the particular physical and chemical composition of our planet, and therefore that the universe must have had a Designer. The argument is very weak. It assumes that there can be no life except in the forms in which we know it, but there seems no intrinsic reason why life of some sort should not exist in a world of very different physical and chemical make-up from our own. Even in our world, the feats of some of the bacteria, living without oxygen for instance, are remarkable. The occurrence of particular temperatures or chemical compounds in our world provides in itself no proof, or even probability, of divine planning.

Darwinism, in the minds of many, eliminated purpose in any form and implied that animal evolution, including that of man, was the result of 'blind chance'. 'The Darwinian process may be described as a chapter of accidents,' wrote Bernard Shaw (in the Preface to *Back to Methusaleh*, 1921) and 'when its whole significance dawns on you, your heart sinks into a heap of sand within you. There is a hideous fatalism about it.' This was the main reason why he and other atheists advocated a Life Force, and various Christians have felt similarly. Canon Smethurst, for instance, supposed that Neo-Darwinians 'attribute the results of the evolutionary process entirely to the blind and fortuitous working of natural selection upon variations produced by mechanistic forces' (*Modern Science and Christian Beliefs*, 1955).

But the use of such words as 'chance', 'fortuitous' and 'accident' in this context is ambiguous and misleading; T. H. Huxley long ago called this 'the most singular of these, perhaps immortal, fallacies' in regard to the theory of evolution (*Life of Darwin*, 1887). As already stressed, mutations are random in relation to the needs of the animal, but natural selection is not. Selection, as the word implies, is the reverse of chance. On a short-term view, the rigour of natural selection is best shown by the relatively uniform appearance of each individual of the same kind of animal, despite the repeated disadvantageous mutations that arise, and also by the speed with which an occasional favourable mutation spreads. On a long-term view, the best evidence that evolution is not random is provided by convergent adaptation, that is, the evolution of different types of animal into similar-looking forms, through their becoming adapted to a similar way of life. For instance, the Galapagos Islands in the Pacific Ocean have been so isolated that most American song-birds have not colonized them. In their absence, one of the few forms which did become established has evolved into a group of species which in their habits and appearance resemble the seed-eating finches, fruit-eaters, insect-eating warblers and tits, and tree-climbing woodpeckers of the American mainland (D. Lack, *Darwin's Finches*, 1947). Similarly Australia was colonized by marsupial mammals which, in the absence of placental forms, evolved into fox-like, wolf-like, mole-like, squirrel-like, rabbit-like, rat-like, anteater-like, and flying-squirrel-like forms, which resemble, often closely, their counterparts among the placental mammals of other continents. Much the same happened during the period in which South America was an island, both marsupial and primitive placental mammals there evolving into diverse and remarkable types, nearly all of which later became extinct, when a land connection was established with Central America which allowed beasts of prey and other newer types of placental mammals to enter from the north (G. G. Simpson, *The Meaning of Evolution*, 1951). Such instances show that evolution has not proceeded at random, and they are fully explicable in terms of natural selection (*cf.* Darwin's Finches, *loc. cit.*); though, owing to the long time involved, they provide no such direct evidence for its action as does a modern example, the evolution of melanic forms in urban areas by several unrelated species of moths (see p. 45, and Kettlewell's paper).

Darwinism has also been attacked for the opposite reason, that so far from being random, it implies a rigid determinism in the course of evolution. Such fears have been vented by both Christians and atheists, but the arguments have been highly confused. Thus on the whole, deterministic views have prevailed among vitalistic rather than materialistic biologists, among those who see a purpose behind evolution rather than among those who see none. This seems logical, since those who deem evolution purposeful cannot admit pure chance. But strict determinism was also upheld by T. H. Huxley: 'The existing world lay potentially in the cosmic vapour and a sufficient intelligence could, from a knowledge of the properties of the molecules of that vapour, have predicted, say the state of the British fauna in 1869, with as much certainty as one can say what will happen to the vapour of the breath on a cold winter's day' (*Life of Darwin*, 1887).

But Huxley's claim is not certain. The temperature at which vapour freezes is a scientific fact capable of repeated verification. As such, it does not seem comparable with the particular sequence of events that constitutes the course of animal evolution, which is more in the nature of an historical fact. Thus a leading modern worker has written of the evolutionary process: 'We can imagine ourselves able to foresee all its possible forms, and to state in advance the probability that each will occur. We can no longer imagine ourselves capable of foreseeing just which of them will occur' (R. A. Fisher, *Creative Aspects of Natural Law*, 1950). G. G. Simpson (*Sci. Monthly*, 71 : 262–7, 1950) has argued similarly, pointing out that while evolution proceeds by general laws, their rigidity (notably in the case of orthogenesis) has been much exaggerated, and that although animals often show convergent trends, each species differs from all others, because it starts with a different hereditary make-up and in a different natural setting. 'The peculiarity of evolutionary determinism consists in its being historical and not mechanistic and in its permitting multiple solutions and not only a unique outcome.'

A deeper discussion would involve considering the true meaning of, and the relationships between, causation, natural laws, and historical sequences of events, problems in philosophy which lie outside my title and beyond my competence. Moreover, their implications for Christian belief are far from clear, and have been variously interpreted. The essential point, as I see it, is that while Darwinism showed animals to have evolved in accordance with natural laws, this observation throws no special light on the question of determinism; hence determinism need not have been made an issue between Darwinists and Christians, and its implications for both sides are highly uncertain.

Behind the criticism that Darwinism means that evolution is either random or rigidly determined lies the fear that evolution proceeds blindly, and not in accordance with a divine plan. This is another problem that really lies outside the terms of reference of biology. It is true that various biologists have inferred that, because evolution occurs by natural selection, there is no divine plan; but they are being as illogical as those theologians whom they rightly criticize for inferring that, because there is a divine plan, evolution cannot be the result of natural selection. The latter point of view is what Wilberforce attacked as 'lying for God' (see p. 15). In science, for the Christian as much as for the research worker, 'moral or emotional grounds for preferring one conclusion to another are completely out of place. Let our loyalty to the facts be absolute' (R. A. Fisher, *Creative Aspects of Natural Law*, 1950). It was this attitude that enabled Charles Kingsley to welcome Darwin's theory from the start, since for him: 'science is the Voice of God—her facts, His words—to which we must each and all reply, "Speak Lord, for Thy servant heareth".' (*Life*, 1883.) The biologist, equally, may agree with T. H. Huxley that 'the doctrine of evolution is neither Anti-theistic nor Theistic. It simply has no more to do with Theism than the first book of Euclid has' (*Life of Darwin*, 1887). Biological discoveries have corrected certain naive and mistaken views as to the manner of creation, and have shown that evolution proceeds by natural laws, but they do not, and cannot, provide any information as to whether natural laws are or are not divinely instituted.

The fundamental argument from design, much more powerful than Paley's, is that a universe which includes the regularities that we call scientific laws, laws which have resulted in nebulae, electrons and intelligent animals, must have had a Planner. This view has been held by many in the past, and by modern scientists such as the late E. A. Milne (*Modern Cosmology and the Christian Idea of God*, 1952) and P. A. Moody (*Introduction to Evolution*, 1953).

For many others, however, the existence of regular laws as certainly implies a materialistic and atheistic universe, a view which F. Temple thought it specially necessary to combat in his sermon before the scientists at Oxford in 1860, quoted earlier in this chapter. Some scientists have agreed with the well-known hymn:

The spacious firmament on high,
With all the blue ethereal sky,
And spangled heavens, a shining frame,
Their great Original proclaim.

J. ADDISON (1672–1719)

Others have concurred with a later poet:

I find no hint throughout the Universe
Of good or ill, of blessings or of curse;
I find alone Necessity Supreme.

J. THOMSON, *The City of Dreadful Night* (1880)

From the cleavage of opinion it may be inferred that powerful reasons can be advanced by both sides, or perhaps that the argument is not rational; certainly, it is not scientific. The evidence of science on this point is neutral, favouring neither theism nor atheism, an opinion that various Christians and various agnostics seem to find shocking, though for opposite reasons.

But any man that walks the mead,
In bud or blade, or bloom, may find,
According as his humours lead,
A meaning suited to his mind.

ALFRED LORD TENNYSON
The Day-Dream (1842)

Tennyson later wrote: 'I found Him not in world or sun, Or eagle's wing, or insect's eye' (*In Memoriam*, 1850); while in a sermon of 1839, J. H. Newman said: 'It is indeed a great question whether Atheism is not as philosophically consistent with the phenomena of the physical world, taken by themselves, as the doctrine of a creative and governing Power' (*Fifteen Sermons preached before the University of Oxford*).

Addison's hymn cited earlier, like the 19th Psalm on which it was based, was concerned not only with the laws but with the beauty of the Grand Design. 'He who stops at the facts misses the glory' (C. A. Coulson, *Science and Christian Belief*, 1955), and for many people it is the grandeur rather than the planning of nature that particularly declares the glory of God. Nevertheless, there are many others who draw no such conclusion, nor need they be persons blind to natural beauty. Hence this argument, though perhaps stronger than the other, is also inconclusive. 'If the heavens do not declare to you the glory of God nor the firmament show you His handiwork, then our poor arguments about them will not show it' (Charles Kingsley, *Scientific Lectures and Essays*, 1880). It may be added in parenthesis that the many biologists who discuss their findings solely in scientific terms are not necessarily blind to the beauty or wonder of what they are studying. Often, their awareness of these attributes has been an incentive in their choice of career, but they properly keep the subject out of their biological writings, since it is not one with which science is concerned. A quotation may be added from T. H. Huxley: 'Whoso clearly appreciates all that is implied in the falling of a stone can have no difficulty about any doctrine simply on account of its marvellousness' (*Life*, 1903). A further difficulty, that nature includes not only grandeur but terror, is considered in the next chapter.

To sum up the different points made in this chapter: The adaptations of animals are due to natural selection and do not, in themselves, constitute evidence for divine creation. Nor should such evidence be sought either in the gaps in existing knowledge of the evolutionary process, or in the particular composition of the inorganic world. That evolution has occurred by natural selection means that it has not been 'random', but is the result of natural laws. Whether or not these laws have rigidly determined the course of evolution, and whether or not this course has been divinely planned, are metaphysical questions outside the scope of biology. Hence the causes of conflict between Darwinism and Christianity discussed in this chapter have been due to unscientific and unwarranted claims, made especially by certain (but not all) Christians in relation to biology, and by certain (but not all) evolutionary biologists in relation to metaphysics.

7

Death in Nature

Various writers, some Christian and others agnostic, have been troubled about natural selection not only because it seems too random (a misunderstanding now, I hope, removed), but also because it is so unpleasant. Natural selection works because in each kind of animal most individuals die before they have produced any offspring, while most of the rest die before they have borne as many offspring as they might. In many wild birds, for instance, over three-quarters of the young die before breeding, and between one-third and one-half of the adults die each year. Hence nearly all die in youth or the prime of life, and an extremely small proportion reaches old age. In such animals as fish and insects, the losses among eggs and young are much higher, commonly exceeding 999 in every 1,000 before maturity is reached (D. Lack, *The Natural Regulation of Animal Numbers*, 1954).

At the start of the last century, Paley (*Natural Theology*, 1802) could write: 'It is a happy world after all. The air, the earth, the water, teem with delighted existence. In a spring noon or a summer evening, on whichever side I turn my eyes, myriads of happy beings crowd upon my view. "The insect youth are on the wing." Swarms of new-born flies are trying their pinions in the air. Their sportive motions, their wanton mazes, their gratuitous activity, their continual change of place without use or purpose, testify their joy and the exultation which they feel in their lately discovered faculties.' But the modern observer sees the swallows and flycatchers preying on this multitude, whose activities, so far from being 'gratuitous', are adapted by natural selection for the attraction of mates or the avoidance of enemies.

Tennyson's *In Memoriam*, published, it may be recalled, in 1850, nine years before *The Origin of Species*, revealed a different picture:

Are God and Nature then at strife,
That Nature lends such evil dreams?
So careful of the type she seems,
So careless of the single life;

That I, considering everywhere
Her secret meaning in her deeds,
And finding that of fifty seeds
She often brings but one to bear,

I falter where I firmly trod . . .

'*So careful of the type*'? *but no.*
From scarped cliff and quarried stone
She cries, '*A thousand types are gone;*
I care for nothing, all shall go.'

In the same poem comes the well-known line about 'Nature red in tooth and claw'. In similar vein was Bernard Shaw's definition of natural selection in his Preface to *Back to Methusaleh* (1921): 'To modify all things by blindly starving and murdering everything that is not lucky enough to survive in the universal struggle for hogwash.' To a percipient atheist like Shaw, it was terrifying that the world should be of such a pattern, and inconceivable that man with his ideals should have evolved by such ghastly means. To various Christians also, the apparent waste and cruelty of natural selection have seemed incompatible with the God of mercy and love.

One strong reason for postulating a Life Force has been this abhorrence of natural selection, and earnest persons, many of them Christians, have sought in nature for some mitigation of the harshness. Canon Raven justly remarked that 'to select the earthquakes or the parasites is as mistaken and almost as irritating as to select the sunsets and the lilies' (*Science, Religion and the Future*, 1943); but while it is true that various instances of co-operation among animals can be found (after diligent search) to set against the many instances of competition, predation, and parasitism, the fact remains that there is a vast annual loss of life. One should not, of course, suppose that the animals suffer terror or pain equivalent to what men suffer under similar circumstances, for so much of terror and pain depends on the higher mental faculties, but at the lowest level of interpretation, there remains a huge seeming wastage of life on the earth.

Nor would it help appreciably if natural selection could be rejected as the agent of evolution, for a huge annual destruction occurs anyway, and on the moral plane it is quite secondary whether or not it is selective. Indeed, many would prefer to think that this destruction has been the cause of evolution rather than that it is purposeless; but either way, this is irrelevant. The only sound attitude is submission to the facts, and Christians may recall Richard

Hooker's words (in another context) that 'in matters which concern the actions of God, the most dutiful way on our part is to search what God hath done, and with meekness to admire that, rather than to dispute what He in congruity of reason ought to do. . . . When we do otherwise, surely we exceed our bounds' (*Of the Laws of Ecclesiastical Polity*, 1592–4). To the same effect, Dr. E. B. Pusey wrote (though in an anti-Darwinian sermon): 'What are we, that we should object to any mode of creation, as unbefitting our Creator?' (*Un-science, not Science, adverse to Faith*, 1878).

The same answer is given in the *Book of Job* to the much harder problem of human suffering. After Job had heard the voice of God, rational complainings became insignificant, and the only possible reply: 'But now mine eye seeth Thee. Wherefore I abhor myself and repent in dust and ashes' (*Job*, 42 : 5, 6). This constitutes only an acknowledgement, not an explanation, but for Christians, the problem of suffering is developed further by example, in the New Testament, a subject that comes outside the scope of this book. It may be added, however, that St. Paul wrote that 'the whole creation groaneth and travaileth in pain together until now' (*Romans*, 8 : 22).

Returning to the problem of death in animals, various Christians have supposed that it could not have been intended in the divine plan, but is the work of the devil, or whatever name is given to the power of evil in the world (see particularly A. F. Smethurst, *Modern Science and Christian Beliefs*, 1955, also N. P. Williams, *The Ideas of the Fall and of Original Sin*, 1927). This seems another instance of 'argument from design' with its attendant weaknesses. Some writers, though not those cited, have based the claim on passages in the Old Testament, notably *Isaiah*, 11 : 6, 7 and *Hosea*, 2 : 18, but these may be disregarded, since Isaiah and Hosea may well have intended their statements allegorically, and if not, they cannot be treated seriously in the light of modern knowledge. The lion could not 'eat straw like the ox' and remain a lion. Others have claimed that nature was corrupted by the Fall. But the fossil record shows that animals died for millions of years before man appeared, while if the Fall is an allegory, it is concerned with man's disobedience to God, and can hardly be connected with natural death in wild animals.

The serious claim is that the power of evil which helped to produce man's Fall was active in the world before man's appearance and produced other effects, including death. But can anyone be confident that the death of animals is evil? Death is essential to the maintenance of the organic world as we know it, for three reasons. In the first place, once any species of plant or animal has reached a stable population, and most of them quickly do so, all births must be balanced by a corresponding number of deaths. A world in which no animals died would have to be one in which no more were born; by implication, it would be a world without reproductive behaviour, with no courtship and with little or no bird song. Secondly, it has been supposed by various writers, following Isaiah, that the animals in such a world would eat plants; but some of the essential nutrients needed by plants are derived from the bacterial decay of dead bodies, and if the plant-eating animals did not die, plant life could not long persist. Thirdly, the only method of evolution of which we have knowledge is natural selection, which requires a high death-rate. These points, so obvious to the biologist, show that the death of animals is inextricably bound up with the continued existence of life on the earth. I would therefore suggest that the Christians who claim natural death to be evil have not realized the implications, and that this view cannot be justified on biological nor, I would suppose, on theological grounds. It should be added that in recent times only a few Christians have expressed such views.

One illustration may be added to show the danger of reading human values into nature. In the glorious weather of May 1940, I was shown a shallow pond on Dartmoor filled with that rare and delicate creature the Fairy Shrimp. In the hot and dry weather, the pond had shrunk and a few Fairy Shrimps had been caught in the drying mud at the sides. By next morning more water had gone, more Fairy Shrimps had died, and the rest were crowded into a dense mass. By the evening of another day, the whole pond had dried up and every Fairy Shrimp was dead. It seemed a pointless tragedy; but inquiry showed that it happens each summer, and that only because it does so can Fairy Shrimps continue to be there. For they fall easy prey to other pond animals, hence cannot survive in permanent waters. But provided that their pond dries out for part of each year, their enemies cannot survive there, whereas Fairy Shrimps lay eggs adapted to withstand desiccation, which hatch out in the following year. It would be wrong to press this instance too far, but it may at least serve as a warning against too superficial a reading of nature's ways.

I have met Christians who, while acknowledging the necessity for animals to die, have yet claimed disease to be evil. But of the three main causes that keep natural populations in check, disease seems no more unpleasant than starvation or predation, and in fact the three commonly work in conjunction. Thus a diseased animal is more likely than a healthy one to be caught by a beast of prey. The parasitic organisms which cause disease show adaptations as wonderful as those of other animals, and we are surely exceeding our bounds in judging them to be evil. This would seem a misapplication of man's moral standards, which, if they come from God, were presumably given as a guide for human conduct, and not for condemning other parts of the creation.

Most human populations have lived under conditions of mortality very similar to those of the beasts. This may have been overlooked by those who write and read books, who are among the minute fraction of our species to which such conditions have not applied, or have applied more mildly than usual. The skeletons of Stone-Age men, the inscriptions on tombs in Ancient Rome, and the records of modern India, reveal that most men, like most animals, have died in childhood or in their prime (D. Lack, *The Natural Regulation of Animal Numbers*, 1954). Some Christians have claimed that this fate was not intended, citing Genesis 3 : 19 that the death of man came as a result of the Fall (*cf.* St. Paul, *1 Corinthians*, 15 : 20.) Such a view was formerly widespread, and is still maintained by Roman Catholics as a part of the Fall story that is not wholly allegorical (see p. 37). But it is hard to see why this particular point should not be treated allegorically, and there seems good precedent for doing so, since according to Gore (*Lux Mundi*, 1891, 12th ed., app. 2), some of the early fathers of the Church, including Athanasius and Augustine, held that spiritual and not bodily death was meant by the verse in question.

That, but for the Fall, man was intended for a different spiritual fate is an essential part of Christian belief. But since man evolved from beasts, it is reasonable to suppose that he would be subject to similar physical limitations while living in this world. This means that, so long as men need food and have children, bodily death is as essential to the continuance of our species as of any other on the earth. Further, the three chief means of regulating numbers, by starvation, predation and disease, apply as much to man as to other animals. It seems illogical to try to exclude disease from the trio, though a fourth means, internecine strife, plays a far larger part in man than any other animal, and might perhaps be avoidable. It can, of course, be argued that if the Fall had not occurred, man's physical nature was to be miraculously changed, so that he was to become immortal and no longer subject to biological limitations, but that is to take the problem outside science.

It should be added that many Christians have not been troubled by the problems considered in this chapter. The same has applied to most atheists and agnostics, except for those who, like Shaw, have found it repugnant to suppose that man's moral and aesthetic standards could have been evolved through natural selection. Summing up, the validity of the theory of natural selection in animal evolution can be determined only on scientific, not metaphysical, grounds. Further, although on theological grounds the ordering of the animal creation may to some persons seem surprising, man is surely unqualified to judge whether this ordering is in any way evil, or contrary to divine plan.

8

The Nature of Man

The most important cause of conflict between Darwinism and Christianity concerns the nature of man. Christians believe that he has a spiritual nature, and a soul given by God, while biologists hold that he has evolved from the beasts by natural selection. Are these views incompatible? How, in any case, did man's unique or almost unique attributes arise, including his self-awareness, his apparent free-will, his claim to distinguish good from evil, and to act or fail to act on this distinction, his capacity by reasoning to reach truth, including abstract truth, his sense of the holy and beautiful, and his idea of God?

The biologist's approach to this group of problems has often been condemned by professional philosophers as crude and naive, while the philosopher's approach has often been dismissed by biologists as having too little to do with facts, as being a play with words and abstract notions, and as being too involved to understand. I speak as a biologist, so it may be worth trying to see how the biologist's attitude is likely to be formed.

First, in his research the biologist is continually bumping up against unexpected facts, and he then searches for repetitions of these facts to establish principles, after which he examines further facts, or makes experiments, to test his principles. From experience, he has come to distrust speculative ideas which are more than a small step ahead of the facts, because later discoveries have so often shown such speculations to be wide of the truth. Frequently, too, new facts precede new ideas rather than the other way about. This method of advancing knowledge is very different from that of the philosopher, it tends to make the biologist unsympathetic towards philosophy and to argue badly when he tries to present a philosophical view.

The contrasting attitudes are well illustrated in Darwin, the naturalist, and T. H. Huxley, who was at least half a philosopher. On reading *The Origin of Species* for the first time, Huxley reflected: 'How extremely stupid not to have thought of that!' (*Life*, 1903). But it needed a personal visit to the extraordinary Galapagos Islands before so new an idea could be driven into a human mind, and even then recognition was far from sudden, and came only after Darwin had pondered on his results for several years after his return to England. Hence the discovery was made slowly by the practical biologist, Darwin, and not in a flash by the more brilliant and quick-thinking Huxley, working it out from first principles. Moreover, after completing *The Origin of Species*, Darwin settled down to collect new facts, about earthworms, orchids and many other aspects of natural history, none of which led again to so great an idea, while T. H. Huxley spoke and wrote in favour of Darwinism and thought out its consequences for mankind. This comment is not at all intended as a denigration of Huxley, but to show that his greatness lay in a different direction from that of Darwin. Both men were intellectual giants, and both played an extremely important part in the promulgation of Darwinism, but their parts were different. Moreover Huxley was also a great anatomist, and it was the scientist in him rather than the philosopher who enjoyed 'the slaying of a beautiful hypothesis by an ugly fact' (*Life*, 1903).[1]

A second point contributing to the make-up of the biologist is that his research depends on repeated observations. It need not, as popularly supposed, consist solely, or even mainly, of measurements and experiments, but unless events are repeated, they cannot be assessed by science. Hence truly unique events come outside the domain of science, though biologists are not usually convinced when told that they must, therefore, leave such problems as miracles to others. For one of the chief ways in which research has advanced is through the discovery of apparent exceptions to the known rules, and if further study shows the exceptions to be replicable, new regularities are revealed from which modified rules can be propounded. This method has been so successful that the biologist tends to doubt whether there are any types of irregularity, or seeming irregularity, that will not yield to it. By some thinkers, human history has been interpreted in a similar way, but most hold that historical events are in an important sense unique, and that historical evidence is of a different nature from scientific evidence, but equally valid in its proper sphere (see also p. 67). The scientist, however, distrusts second-hand evidence and distrusts statements that have to be accepted on authority and cannot be verified by further observation or experiment, an attitude of scepticism that has been fully justified in the progress of science.

Thirdly, most biologists reject any pseudo-mystical factors

[1] Lt.-Col. Meinertzhagen was twice taken as a child to tea with Huxley and Darwin. 'What struck me most about Huxley and Darwin was the anxious, worried, lined face of the former and the serene, unworried face of Darwin.' (*in litt.* to the author).

brought in to account for what cannot yet be explained, finding that a Life Force or similar agent does not help to advance knowledge and that the apparent need for it vanishes when further discoveries have been made. It should be added, however, that biologists have not always been careful in their own use of terms. Such words as 'drive' or 'instinct' in behaviour, and 'organizer' in embryology, are theoretical concepts which have to be treated carefully or they come perilously near to mythology.

Fourthly, the biologist tends to choose the simplest of the available hypothesis until it can be shown to be inadequate. To take only one example, in the Galapagos Islands one of the native finches inserts a stick into cracks in wood, thereby probing out insects for its food (D. Lack, *Darwin's Finches*, 1947). When the ordinary person hears of this use of a tool he supposes that it denotes intelligence, but the biologist supposes that it is instinctive, like most other behaviour in birds, unless additional facts exclude this simpler view. It is generally agreed that such economy of hypothesis is the most effective method for science. It means that the biologist will try to interpret man's peculiar attributes in the simplest possible way or, as some would say, at the lowest possible level.

Now these points in the biologist's attitude, his slow and factual, rather than speculative, advance, his fitting of observed irregularities into new rules, his distrust of second-hand evidence and undefinable agents, and his choice of the simplest interpretation, are not merely principles learnt from teachers or read in books. They are deeply ingrained through practice, he lives and works with them, often unconsciously, and they lead to success in research. Added to this, the biologist of today has been brought up on the idea that the conflict between Darwinism and Christianity has been a steady retreat of religion before science. So many apparent strongpoints of religion have been made untenable by scientific discoveries that he tends to assume that, in the end, all of man's nature will have been explained in scientific terms, leaving philosophy and theology with nothing.

With this background, the biologist approaches the remaining stronghold, man's spiritual, or allegedly spiritual, nature. His equipment, he may agree, is as yet far too meagre for an attack to be successful, but even though he ignores the traditional routes of advance laid down by philosophy, he is confident of eventual success. The spectacle is ennobling or fatuous, according to taste; and because the challenge has been issued by an army triumphant from earlier victories, many now believe that the war is as good as won, and that man has no spiritual nature.

In the initial enthusiasm for Darwinism, biologists stressed man's affinities with beasts. Thus when T. H. Huxley wrote *Evidence as to Man's Place in Nature* in 1863 he was primarily concerned to show that the anatomy of man strongly resembled that of the higher apes, and that he differed in no essential structure from other animals. It was left to his grandson, Julian Huxley, to redress the balance, and to point out that *Homo sapiens* has mental qualities which separate him by an immense gulf from other living animals (*The Uniqueness of Man*, 1941).

The Victorians' fear that Darwinism made man merely an ape was partly, as pointed out by Canon Liddon, a misplaced family pride (*The Recovery of St. Thomas, with a prefatory note on the late Mr. Darwin*, 1882); and as Julian Huxley has stressed (*Evolution and Ethics*, 1947), because man arose from apes, it does not follow that he is 'nothing but' an ape. Darwin, likewise, asked in *The Descent of Man* (1871), why should it be 'more irreligious to explain the origin of man as a distinct species by descent from some lower form, through the laws of variation and natural selection, than to explain the birth of the individual through the laws of ordinary reproduction?' 'Both,' Darwin added, 'are equally parts of that grand sequence of events, which our minds refuse to accept as the result of blind chance.'

In *The Descent of Man*, Darwin cautiously argued that, since man was a product of natural selection, his mental or spiritual qualities should be derivable from rudiments present in other animals, which rudiments he then described in so far as they were known to him. Darwin's evidence was largely anecdotal, but modern studies of animal behaviour have confirmed his view that at least some of man's special mental attributes occur in rudimentary form in other mammals. Learning and intelligence do so, and though specially characteristic of apes, have also been demonstrated in birds, fish and the main invertebrate groups. Birds, for instance, have a sense of number (W. H. Thorpe, *Learning and Instinct in Animals*, 1956). On the other hand, research on the innate or instinctive behaviour of animals suggests that man himself may have a much larger instinctive component in his make-up than was at one time supposed. This will doubtless become clearer when objective studies, like those of N. Tinbergen on animals (*The Study of Instinct*, 1951), come to be applied to man, to supplement the intuitions, often based on subjective data, of the psychoanalysts.

The emotions of animals are much harder to assess, but

the overt behaviour of many mammals and of some other animals suggests that they possess traces of feelings present in man. Thus in mammals the family unit, usually a mother and young, seems bound by strong emotional ties. The extent of such ties among broader social units is much more doubtful, but they may be present to some degree in apes. Dogs are most often cited in this respect, for they show apparent pleasure in our company, loyalty in defence of our belongings and persons, seeming guilt at misdoing, and other human traits. Some thinkers have argued that such traits have been acquired by dogs through their association with man, but what is now known of animal behaviour makes it unlikely that dogs could respond to man in such ways unless their behaviour had a strong innate basis, presumably evolved through natural selection, in their original group-life in the pack or other social unit (K. Lorenz, *Man meets Dog*, 1954). On the subjective plane, of course, it is impossible to evaluate the emotional life of animals, and the question of whether they have conscious awareness and, if so, what this statement might mean, cannot be answered. For the same reason it is impossible to say whether animals possess a sense of beauty. Some of the most beautiful actions in nature, to our human senses, are the songs and displays of birds. These seem largely innate or instinctive, and it may be doubted whether the birds themselves perceive them as beautiful, even in the case of the Satin Bower-bird, which decorates a display area with blue and yellow objects, and also 'paints' the surrounding vegetation with the help of berries (A. J. Marshall, *Bower-birds*, 1954).

The two mental qualities of man usually regarded as the most important are his power of abstract reasoning and his moral sense, and there seems a provisional case for thinking that rudiments of these may be present in other animals, though certainty is difficult. Darwin argued further that man's conscience might have arisen in the course of evolution through a combination of social instincts and reasoning, but this idea is more doubtful, being linked with problems of ethics and values considered in the next chapter. Darwin did not much concern himself with philosophy, his chief aim being to show that 'the difference in mind between man and the higher animals, great as it is, certainly is one of degree and not of kind', and hence that man's possession of special mental attributes provided no valid objection to his having evolved from animals (*The Descent of Man*, 1871).

One remaining human attribute, of which no trace is found in animals so far as known, is the sense of the holy, and the desire, however naively expressed, to worship and sacrifice to a Being higher than man. Some have argued that this is more basic than man's moral sense, but others have claimed it to be a compound of other feelings and actions, derived especially from the infant's emotional relationship to its parents, and of no particular significance.

This leads on to the most serious difficulty for the Christian in relation to human evolution, that biologists hold man to have evolved gradually over many generations from apelike ancestors, while Christians hold that he differs absolutely from animals in possessing a soul. This is the remaining problem, deferred from Chapter 3, that is raised by the question of whether Adam and Eve were historical beings. As stressed by Canon Liddon (*The Recovery of St. Thomas, with a prefatory Note on the late Mr. Darwin*, 1882). 'If it should ever become certain that the first man had for his mother an anthropomorphous ape, the Church's faith as to man's true place among the creatures of God would remain untouched. A "separate creation" must have invested this Adam before his birth with that living soul in whose faculties lay his likeness to the Almighty Creator. No evolution can have led up to this great gift.' Liddon added that this truth does not belong to the fringe of Christian belief, it is of its essence. The same point has been emphasized repeatedly by theologians in the course of the Darwinian controversy, and once again in the papal encyclical *Humani generis* of 1950 (as quoted on p. 35).

The soul is regarded by Christians as the spiritual part of man, though it is closely linked with his body; it is bound up with the essential personality of each individual, and each soul is unique; it is also responsible for moral decisions and rational conclusions, and it is immortal. It is hard to give a more precise definition, and some have regarded the soul as an unnecessary concept. Perhaps it means no more than the individual personality, with the nature and attributes that Christians ascribe to it, but as such, the term is convenient. Differences of opinion among Christians concerning the origin and nature of the soul, and some of its possible characteristics, come outside the scope of this book. Indeed, it might have been thought that, since the soul is held to be spiritual, the question of whether it exists would lie outside scientific inquiry, and so would raise no problems in relation to Darwinism.

Actually, two serious difficulties are involved. The first arises because, according to Christian theologians, animals other than man do not have souls. With the theory of evolution, the old idea of the special creation of each form of life was replaced by a concept of continuous change from preexisting forms. Darwinists such as T. H. Huxley rightly attached much weight to the principle of continuity in evolution, but this principle is broken if, as Christians hold, man differs absolutely from other animals in one particular. even if the attribute in question is spiritual. This consideration has led many to reject the concept of the soul and to suppose that man has no spiritual element in his make-up.

The claim is sometimes made that a soul, or spiritual element, is present in rudimentary form in other animals, and hence that the human soul has been gradually evolved. This would preserve the principle of continuity in evolution. It seems inconceivable, however, that a soul could be evolved by natural selection, hence this view is usually coupled with belief in emergent evolution or a Life Force, which to biologists is inadmissible; and since the claim that animals have souls is also rejected by orthodox Christians on theological grounds, the whole idea may, I think be dismissed. At least, it plays no part in the conflict between Darwinism and Christianity.

Either man, like other animals, has no spiritual part to his nature, or, as on the Christian view, a supernatural event took place at the time of man's first appearance, before which our ancestors were proto-human mammals and after which, through the divine gift of a soul, they were truly human. This certainly means a break in evolutionary continuity, but it does not mean a return to the view, rejected earlier, that the natural means of evolution have been supplemented by miracles, since no change in, or suspension of, natural laws is held to have been involved. 'The creation of the first human soul was an event outside the order of physical causation, but it was no more miraculous, in the strict sense of the term, than is the creation of any other human soul' (J. L. Russell, *Month*, *n.s.*, 15:

33–45, 1956). Since the change in question is held to be outside the province of science, the Christian view is compatible with the theory of evolution, but only provided that the attributes regarded as peculiarly human, and made possible by the change, are considered to be spiritual, and hence outside biology.

This brings up the second and more serious difficulty, since these special human attributes, of which moral responsibility is perhaps chief, are inevitably held by Darwinists to have evolved by natural means. The possible ways in which this might have happened are considered in the next two chapters, and further questions concerning whether man has or has not a spiritual nature may be deferred until these views have been examined.

9

Evolutionary Ethics

The theory that man's moral standards and ethical ideas have been evolved from the behaviour of other animals by natural means is usually called 'evolutionary ethics', a term that I have kept, though the word 'evolutionary' in this context is rather misleading. Because man evolved from other animals, it seems reasonable to infer that his moral sense evolved too. So Darwin argued in *The Descent of Man* (1871), though he ascribed moral standards not to natural selection but to human reasoning. The first advocate of true evolutionary ethics was Herbert Spencer, who was a railway engineer, then a journalist, and eventually the 'philosopher of evolution'. As such, he introduced the phrase 'survival of the fittest', which for a long time replaced Darwin's more accurate and less tendentious 'natural selection'. He was an agnostic, and a great personal friend of T. H. Huxley. Some of his books had a popular vogue, but they are now rarely read, and the high repute in which he was held in his lifetime has not been confirmed by posterity.

There had, of course, been utilitarian or rationalistic theories of ethics before that of Herbert Spencer, who combined some of these views with his own. Thus he supposed that happiness was the goal to be attained and that the evolutionary process brought man there. He also held that good is 'the conduct which subserves life, and bad the conduct which hinders or destroys it', that 'the highest form of conduct' makes 'the totality of life greater', and that 'the conduct to which we apply the name good is the relatively more evolved conduct; and bad is the name we apply to conduct which is relatively less evolved' (*The Data of Ethics*, 1879). Hence he equated that which was evolved later with that which promotes both life and happiness. As to how ethical ideas might be evolved, Spencer was helped by believing, like many of his contemporaries, that acquired characters could be inherited (see pp. 49-50).

It would seem that many later Darwinists have believed in evolutionary ethics. Thus the zoologist Ray Lankester wrote of Darwinism that 'its most important initial conception is the derivation of man by natural processes from ape-like ancestors, and the consequent derivation of his mental and moral qualities by the operation of the struggle for existence and natural selection from the mental and moral qualities of animals' (quoted from the *Encyclopedia Britannica*, in order to argue against it, by St. G. Mivart in *The Origin of the Human Reason*, 1889). Again, J. H. Woodger thought it necessary to refer to the theory, criticizing it adversely, in his *Biological Principles* (1929). Nevertheless, evolutionary ethics was not treated substantially between Herbert Spencer's and modern times, when the theory has been re-stated by three zoologists of repute, namely C. H. Waddington (*Science and Ethics*, 1942, initiating a discussion to which others contributed), Julian Huxley (in the Romanes Lecture for 1943, reprinted with additions in Huxley, T. H. and J., *Evolution and Ethics*, 1947), and G. G. Simpson (*The Meaning of Evolution*, 1951).

The modern upholders of evolutionary ethics accept the theory of natural selection and hence postulate that high moral standards have been evolved by man because they increase the chances of survival of himself and his offspring. On this view, evil actions seem best explained as the remains of behaviour evolved before man was social, or at a time when he was at a more primitive social stage, behaviour that once helped him to survive but does so no longer. This view, however, is more plausible in relation to 'animal' sins like lust than to those involving the higher faculties, like pride; and following the precepts of the New Testament, Christians hold pride to be more evil than lust.

A much more serious difficulty is to see why natural selection should necessarily lead to higher moral standards. As Darwin pointed out in *The Descent of Man* (1871): 'It is extremely doubtful whether the offspring of the more sympathetic and benevolent parents, or of those who were the most faithful to their comrades, would be reared in greater numbers than the children of selfish and treacherous parents belonging to the same tribe.' Darwin therefore rejected natural selection as the agent of moral improvement, and many will agree with him.

Because of this difficulty, some have argued that higher moral standards help the survival, not of the individuals holding them and of their offspring, but of the societies of which they are members. On this view, selection has been of the group, not the individual, those human societies with higher moral standards tending to supplant those with lower moral standards. Such group-selection is not, of course, natural selection in its ordinary biological meaning, since natural selection operates on hereditary characters in individuals and their offspring. This does not, in itself, constitute an objection to the theory but it means that any analogy with animal evolution is dangerous, especially since the possibility of group-selection at the biological level has not been proved and remains in great dispute. (The social insects do not provide a true parallel, since in bees, for instance, the hereditary characters of the whole colony are borne by the queen.) But the basic criticism of this view is the same as that raised by Darwin in relation to individuals, that there is no valid reason for supposing that those societies with higher moral standards necessarily come to supplant those with lower moral standards.

Another great difficulty concerns the basis for man's standards of good and evil. The upholders of evolutionary ethics have usually stated or implied that what has evolved is good, either necessarily in itself, or because it conforms to some standard which they hold to be the direction of evolution. Herbert Spencer's views on this point were quoted at the start of the chapter. Of the modern advocates, J. Huxley wrote: 'Evolutionary ethics must be based on a combination of a few main principles: that it is right to realize ever new possibilities in evolution, notably those which are valued for their own sake; that it is right both to respect human individuality and to encourage its fullest development; that it is right to construct a mechanism for further social evolution which shall satisfy these prior conditions as fully, efficiently, and as rapidly as possible' (*Evolution and Ethics*, 1947). Later he wrote: 'Anything which permits or promotes open development is right, anything which restricts or frustrates development is wrong. It is a morality of evolutionary direction' (*Evolution in Action*, 1953). Again,

G. G. Simpson criticized earlier writers for holding up an arbitrary external standard to which evolutionary ethics conform (Herbert Spencer's idea that good conduct makes 'the totality of life greater', for instance). He then claimed that man's moral standards should have regard to his possible future rather than his evolutionary past, and announced several 'propositions of evolutionary ethics derived from specifically human evolution', which included the 'promotion of knowledge' and the 'responsibility' which that implies, also acceptance 'of the integrity and dignity of the individual' and promotion of 'the realization or fulfilment of individual capacities', provided that this is done with proper safeguards for the social group (*The Meaning of Evolution*, 1951).

Both J. Huxley and Simpson advocated moral or social aims of which many other people would approve, but their claim that such aims are demonstrated or justified by evolution is quite unconvincing. Simpson's criticism of earlier writers for applying an arbitary external standard by which to judge the good holds equally for himself, and both he and Huxley seem merely to have read their own judgments into evolution (see the criticisms by W. H. Thorpe, *Brit. Soc. Biol. Counc. Occ. Pap.*, 7, 1950 and H. J. Paton, *The Modern Predicament*, 1955). Victorian businessmen are said to have justified laisser faire economics and starvation wages for factory hands by reference to natural selection, a conclusion which Huxley and Simpson would rightly reject, though they themselves used a similar type of argument, since they likewise have held up a part (though a different part) of the evolutionary process as a guide for human actions.

While Spencer, Simpson and J. Huxley have applied

some external and apparently arbitrary standard for judging what is good, such as subserving life, openness of development, or the fulfilment of individual capacities, and have held that this standard was justified by evolution, Waddington's position was simpler and more logical: 'Do we need some external criterion to decide what is the "good" direction of evolution? . . . I think one can answer that no criterion external to the natural world is required.' Again, 'we must accept the direction of evolution as good simply because it *is* good according to any realist definition of that concept.' (*Science and Ethics*, 1942). This is similar to the last quotation given earlier from Herbert Spencer (p. 92), and it would seem the only possible justification for evolutionary ethics, since the moral standard is the evolutionary process itself, and is not outside it. Nevertheless, on this view also, an external standard has been used to assess the good, and Waddington's definition fails to stand, since it is legitimate to hold that what has evolved is bad. Effectively, survival is made the measure of good, which few people would accept. Indeed, rather than let survival determine their moral standards, many men have been ready to die for their beliefs.

Waddington's view presents the further difficulty that men cannot be sure of what is the present or future direction of evolution, so are given no reliable guidance for their actions (especially if their choice should itself be a factor determining the direction of evolution). Finally, many philosophers hold that it is impossible to replace the word 'good' by any other term or description without leaving out something essential, so that the good cannot be judged by any criterion external to it, such as giving happiness to the greatest number of people, or following the direction of evolution (see, for instance, C. E. M. Joad in Waddington's *Science and Ethics*, 1942).

Another great difficulty is to explain man's preference for the good. This was not discussed by Simpson or J. Huxley, but Waddington defined 'ethical principles as actual psychological compulsions derived from the experience of the nature of society'. But 'the moral experience in its authentic form is surely the opposite of a compulsion. The agent believes himself to have the responsibility of choice and the ethical "ought" is recognized not as something which must be obeyed but as something which deserves to be obeyed, though it may be difficult and unpleasant' (W. R. Matthews in Waddington's *Science and Ethics*, 1942). Various other advocates of naturalistic ethics have argued that all men should be fully educated and then, with sufficient knowledge and reasoning power, they will choose what is good. But this, of course, is untrue. The dilemma of conduct is that man, knowing what is good, so often chooses evil. Ovid wrote 'Video meliora proboque, deteriora sequor', St. Paul that 'the evil that I would not, that I do' (*Romans*, 7 : 19); and nearly all will agree with J. B. S. Haldane that science 'cannot, of course, give an answer to the question, "Why should I be good?"' (*Science and Ethics*, 1928).

This brief sketch, by a biologist untrained in philosophical thinking, has I hope been a fair statement and fair criticism of the theories of evolutionary ethics so far put forward. It seems, to me at least, that such theories have failed to account for all the important points, namely man's possession of moral standards and ethical ideas, the nature of those standards, man's preference for the good, and why, knowing what is good, he so often chooses evil. They have failed, in short, to reach the essence of the problem, and have failed very badly indeed. Hitherto, such theories have been advocated in print only by atheists or agnostics, but I have heard Christian biologists speak in favour of evolutionary ethics in discussion, though their views were not fully thought out. I do not see how the theory that moral standards result from natural selection could be truly combined with Christianity, but add this comment in case evolutionary ethics, like the Life Force, is an idea that is liable to change sides in the conflict over Darwinism. It should also be made clear that many other atheists and agnostics reject evolutionary ethics, indeed I have not read any professional modern philosopher who upholds this idea.

T. H. Huxley, the greatest of all Darwinists and also the first agnostic, disagreed with evolutionary ethics so strongly that it broke his long friendship with Herbert Spencer. 'I hear much of the "ethics of evolution". I apprehend that, in the broadest sense of the term "evolution", there neither is, nor can be, any such thing. The notion that the doctrine of evolution can furnish a foundation for morals seems to me to be an illusion which has arisen from the unfortunate ambiguity of the term "fittest" in the formula "survival of the fittest" . . . But the "fittest" which survives in the struggle for existence may be, and often is, the ethically worst.' (*Life*, 1903.) In his Romanes Lecture at Oxford in 1893, given fifty years before that of his grandson already quoted, T. H. Huxley claimed that 'cosmic evolution may teach us how the good and the evil tendencies of man may have come about; but, in itself, it is incompetent to furnish any better reason why what we call good is preferable to what we call evil than we had before.' Ever a fighter, he held that 'the ethical progress of society depends, not on imitating the cosmic process, still less in running away from it, but in combating it' (T. H. and J. Huxley, *Evolution and Ethics*, 1947).

T. H. Huxley did not explain how the cosmic process could, of itself, give rise to a system of ethics that was basically opposed to it, and many have felt this to be an impossible inconsistency in his view. He seems merely to have accepted the moral standards of his day as right, while rejecting the Christian theology on which they were founded, and claiming no other foundation for them. He thus laid himself, and those who think like him, open to the criticism of A. J. Balfour (*The Foundations of Belief*, 1895) that 'their spiritual life is parasitic: it is sheltered by convictions which belong, not to them, but to the society of which they form a part: it is nourished by processess in which they take no share. And when those convictions decay, and those processes come to an end, the alien life which they have maintained can scarce be expected to outlast them.' Many feel that this prediction is being realized at the present day.

Although the existing theories of evolutionary ethics have failed, it might be claimed that this is due merely to insufficient knowledge, and that a satisfactory solution along such lines can be expected later, just as many of the evolutionary problems that puzzled Darwin and his contemporaries have since been solved. Further research, it may be claimed, will reveal new facts concerning the behaviour of other mammals and of human infants, and concerning the parts played respectively by hereditary and nurtural factors in behaviour, and such discoveries will throw fresh light on how moral behaviour develops in the child, and perhaps on how it was acquired by the human race. But none of these considerations seems likely to approach the heart of the matter, which is our seeming power of choice between actions, and why we feel we ought to choose in one way rather than another. Science, the philosophers assert, is concerned only with what is, and not with what ought to be, and if this assessment is correct, no theory of evolutionary ethics can succeed.

10

Alternatives to Evolutionary Ethics

If the theory of evolutionary ethics has to be rejected, i.e. if moral standards were not evolved because they conferred survival value, it might still be true that man has evolved wholly by natural selection, if his moral standards are a product or by-product of some other faculty that has helped him to survive. Such a view, which might be termed indirect evolutionary ethics, has been put forward in two forms, the one attributing morality to reasoning and the other to the subconscious.

Darwin argued in *The Descent of Man* (1871) that while conscience and the moral sense depend in part on social instincts and sympathies, supported by social opinion, they are controlled by the intellect, with its capacity for rational reflection. In the nineteenth century, rationalistic theories of ethics were advocated by several philosophers; but they were strongly criticized by others, the critics arguing that while reasoning helps to clarify moral issues, it cannot be responsible for the ultimate distinction, or choice, between right and wrong. The general argument is outside the scope of this book, but a difficulty from the evolutionary viewpoint is relevant, one that was mentioned by Darwin in his private autobiography (see Frontispiece) and again in a letter of 1881: 'With me the horrid doubt always arises whether the convictions of man's mind, which has been developed from the mind of lower animals, are of any value or at all trustworthy' (*Life*, 1887). But if man's reasoning was evolved solely because it has helped him to survive, the 'horrid doubt' would seem a certainty; hence even if human reasoning were, as Darwin thought, responsible for moral conduct, its conclusions as to the nature of the good could not be trusted, if the good means anything more than that which helps survival.

Two modern upholders of evolutionary ethics, J. Huxley (*Evolution and Ethics*, 1947) and especially C. H. Waddington (*Science and Ethics*, 1942) have taken a different view, supplementing their theories with the Freudian idea that moral attitudes are a product of the subconscious mind, formed primarily during the period when the infant is emotionally dependent on its parents. They did not, however, consider the evolutionary implications of the Freudian concepts. If morality depends on attitudes acquired in infancy, then it is presumably not concerned directly with heritable tendencies, on which alone natural selection operates. Hence it is not at all clear how a Freudian view can be combined with evolutionary ethics. A general evaluation of psycho-analysis would be out of place here, but it may be added that, in so far as it is scientific, it is concerned with what is and not with what ought to be, and so cannot solve the problem of morality.

The theory of evolutionary ethics has been based on the assumptions that man has evolved by natural selection and that high moral standards are valid. If all forms of the theory are mistaken, might it be because one of these underlying assumptions is wrong?

The immense difficulty in supposing that moral and

other human values are a product of natural evolution has been for many people the most cogent argument against natural selection and in favour of a Life Force or holistic urge. The Life Force, it was felt, might be purposive instead of random, and might provide the external standard for goodness that is missing from naturalistic theories of ethics, especially if, as some have held, it emanates from an Absolute. But as has already been made clear, biologists accept natural selection as the means of evolution, and there are strong scientific arguments against any form of Life Force. In the moral sphere, also, it is hard to see how a Life Force would account for the existence of evil, or would provide a reliable guide for human conduct.

If natural selection has to be accepted, should the validity of high moral standards be questioned? It is curious that, while many have written of Darwinism's battle with Christianity, few have considered its undermining of secular humanism. Yet as Bernard Shaw wrote, there is about the Darwinian process 'a ghastly and damnable reduction of beauty and intelligence, of strength and purpose, of honour and aspiration' (Preface to *Back to Methusaleh*, 1921). For if moral standards are really determined through their survival value, any other value is irrelevant, and the reasonable man might well disregard them when they conflict with his personal advantage or that of his children. Since, however, most people have not heeded Shaw's warning, they presumably feel confident that goodness, truth and beauty have real value and importance.

These considerations suggest that the inadequacies of the theories of evolutionary ethics cannot be ascribed to false underlying assumptions concerning either natural selection or high moral standards. Hence the dispute at Oxford in

1860 was not, after all, so simple or one-sided. Faultily though Bishop Wilberforce may have argued, he was aware that natural selection dispensed with moral values (as shown by his review of *The Origin of Species* in *The Quarterly Review* for 1860). T. H. Huxley, on the other hand, accepted moral standards close to the Christian standards, though rejecting their Christian foundation and finding no other basis for them.

As mentioned earlier, scientists have often pictured their conflict with theology and philosophy as a steady advance, with field after field of knowledge brought under their sway, including in recent times much of human nature. On this view, morality is one of the few unconquered strongholds. Another is the capacity to apprehend truth, but Darwin's 'horrid doubt' as to whether the convictions of man's evolved mind could be trusted applies as much to abstract truth as to ethics; and 'evolutionary truth' is at least as suspect as evolutionary ethics. At this point, therefore, it would seem that the armies of science are in danger of destroying their own base. For the scientist must be able to trust the conclusions of his reasoning. Hence he cannot accept the theory that man's mind was evolved wholly by natural selection if this means, as it would appear to do, that the conclusions of the mind depend ultimately on their survival value and not their truth, thus making all scientific theories, including that of natural selection, untrustworthy.

The appreciation of beauty, in both art and nature, raises similar, though less urgent, difficulties. From the scientific viewpoint, also, it is hard to see how free-will or individual responsibility for conduct could be other than illusory, and various scientists, including Freudians, have claimed that they are illusory. But while, formerly, the influence of hereditary, nurtural and economic factors on beliefs and conduct was underestimated, few if any people are prepared in practice to exclude the factor of individual responsibility, and nearly everyone regards it as the ultimate and decisive factor. Finally, there is the problem of man's self-awareness. It seems impossible to formulate this in scientific terms, and hence impossible to study its evolution. Yet it is not merely one part of human experience, but is central, and on it all the rest (including science) depends.

These considerations suggest that an essential part of human experience and human nature lies outside the terms of reference of science. 'Man lives in two worlds (or perhaps more), . . . there is a field, or world, of science in which questions posed in scientific terms get scientific answers; and another world, where words like belief, love, splendour and majesty have meaning. This other world refuses to be shut out of our experience' (C. A. Coulson, *Science and Christian Belief*, 1955). In relation to moral freedom, again, 'if, as seems probable, the scientific point of view is incompatible with freedom . . . then as moral agents we have to maintain that the scientific point of view is not enough. There are two points of view—the moral and the scientific—and while each may be valid within its own sphere, it is from the moral point of view that we get the fullest insight into human action.' (H. J. Paton, *The Modern Predicament*, 1955.) This idea is not, of course, new, but it has needed re-stating in an age dominated by science.

It was stressed earlier that, whatever beliefs a person holds, he should accept the findings of science. It is equally important to ensure that claims made by scientists in the name of science relate to genuinely scientific matters. When such claims are really philosophical, this should be made clear, since otherwise they may acquire a spurious authority. In particular, the claim for which Darwinism stands, that man has evolved wholly by natural means, is a philosophical and not a scientific claim; which has not always been made clear.

If an essential part of human nature lies outside the terms of reference of science, then, since all natural phenomena can be studied scientifically, it would appear to follow that man has not evolved wholly by natural means. If, on the other hand, man evolved wholly by natural means, then all human nature should be interpretable in scientific terms, which would seem to make individual responsibility an illusion, high moral standards a mistake, and intellectual conclusions untrustworthy. Yet individual responsibility, morality and truth seem valid in the light of other, though non-scientific, experience of human nature.

The Christian resolves the difficulty by supposing that man has spiritual characteristics, with which morality and individual responsibility are linked. As set out in allegory in Genesis, God, who is wholly good, created the world, including man, but man sinned, i.e., disobeyed God's will for him. The doctrine of the Fall and of Original Sin seeks to interpret what Christians regard as a fact of human nature, that 'we do not start fair, with a neutral disposition which can, by our own efforts and the help of grace, be converted into one of positive holiness, but with a definite bias towards evil contracted quite apart from any sins of our own' (W. L. Knox and A. R. Vidler, *The Development of Modern Catholicism*, 1933). This bias is held to be due to a subordinate power of evil in the world which influenced the first and all later men. 'We wrestle not against flesh and blood, but against principalities, against powers' (St. Paul, *Ephesians*, 6 : 12). '*If* there be a God, *since* there is a God, the human race is implicated in some terrible aboriginal calamity. It is out of joint with the purposes of its Creator' (J. H. Newman, *Apologia pro Vita sua*, 1864).

The nature of the Fall has been variously interpreted in different ages (see N. P. Williams, *The Ideas of the Fall and of Original Sin*, 1927). Roman Catholics still hold that 'original sin is the result of a sin committed, in actual historical fact, by an individual man named Adam, and it is a quality native to all of us, only because it has been handed down by descent from him' (quoted from the papal encyclical *Human generis*, 1950, transl. R. A. Knox). That is the chief reason why Roman Catholics maintain that Adam and Eve were the parents of the whole human race. Modern Protestants often take a more allegorical view of Genesis, and while acknowledging that all men have a bias towards evil, tend to leave open the question of whether this comes through direct descent from Adam, or in some other and undefined way.

Whether a more literal or a more allegorical view is taken, the doctrine of the Fall is basic to Christian belief. The statement by Darwinists such as G. G. Simpson (*The Meaning of Evolution*, 1951) that 'man has risen, not fallen' misses the point. The human race evolved from beasts, and each adult was once an unconscious embryo, but until a man has attained to his peculiar powers, he cannot misuse them. An act that we regard as evil for a man would not be evil if performed by a bird, since evil comes in question only when there is responsibility for action. The Christian view is superior to that of evolutionary ethics in recognizing the importance of individual responsibility, high moral standards and the tendency for man to do evil; and though the language used is allegorical, not scientific, T. H. Huxley acknowledged that 'it is the secret of the superiority of the best theological teachers to the majority of their opponents that they substantially recognize these realities of things, however strange the forms in which they clothe their conceptions. The doctrines of . . . original sin, of the innate depravity of man . . . of the primacy of Satan in this world . . . of a malevolent Demiurgus subordinate to a benevolent Almighty, who has only lately revealed himself, faulty as they are, appear to me to be vastly nearer the truth than the 'liberal' popular illusions that babies are all born good, and that the example of a corrupt society is responsible for their failure to remain so; that it is given to everybody to reach the ethical ideal if he will only try . . . and other optimistic figments' (*Life*, 1903).

The Christian must accept the findings of science, and hence should accept man's evolution by natural selection. But natural selection is amoral, so cannot have produced man's moral or spiritual characteristics. These might, as suggested in Chapter 8, have been received through supernatural intervention at the transition from proto-human to true man. There is little, if anything, to recommend another suggestion, that moral purpose was somehow embedded in the evolutionary process from the start, as it is hard to attach any precise meaning to it, and it comes dangerously close to the doctrine of the Life Force, if not to pantheism. Other Christians may prefer not to speculate, accepting man's evolution by natural selection and his possession of spiritual attributes, and acknowledging that while these statements seem contradictory, both are in fact true; they belong to different, though complementary, fields of knowledge, which, for the present at least, cannot be discussed in the same terms of reference. St. Augustine held that 'if we cannot resolve such a contradiction [between scientific and religious knowledge] we are to suspend judgment, not doubting either the Holy Scripture or the results of human observation and reasoning, but believing that it is possible, given sufficient knowledge and understanding, to reconcile the apparent contradiction' (quoted by F. S. Taylor, *Man and Matter*, 1951).

This is not the only, or the hardest, contradiction for the Christian. It is also difficult to comprehend how, if God is wholly good, evil could ever have arisen, and how, if God is all-knowing (so knows what is to come), man can have any responsibility for action. Secular humanists may therefore argue that Christian belief should be rejected, but they themselves are in as great a dilemma, though a different one, since it seems impossible to justify high moral standards or abstract truth from the evolutionary process, and individual responsibility seems incompatible with science. If morality, truth and individual responsibility are valid but lie outside science, then Darwinism can never give an adequate account of man's nature. Hence Darwinism is irreconcilable with Christianity, and with at least some forms of secular humanism. Other philosophical systems exist, some theistic and others atheistic, but their consideration falls outside the scope of this book; they seem to involve gaps and contradictions at least as great as those apparent in Christianity on the one hand and secular humanism on the other.

11

The Continuing Conflict

The conflict between Darwinism and Christianity is rarely discussed any more, at least among biologists. Both the many non-Christians and the few Christians among them tend to think, though for different reasons, that it lies in the past. Yet if the arguments in the last chapter are sound, Darwinism seems irreconcilable with Christian, and perhaps with any, moral standards. If there is not enough evidence to solve a problem in science, most people keep an open mind while awaiting further information. But everyone has to take moral decisions, and so forms a basis for his conduct, and this applies as much to those who suppose that moral standards need no basis as to those who postulate a supernatural sanction. It is not possible to keep an open mind about morals, and that is why the conflict between Darwinism and Christianity regarding the nature of man was fought so violently in the past; while the present lack of interest in it might be held to reflect a moral decline. As R. A. Knox pointed out in *Absolute and Abitofhell* (1912), the Victorian 'I believe' was later corrected to 'one does feel'; and now it is 'couldn't care less'.

It is outside the plan of this book, and beyond my competence, to state the general arguments for and against Christianity, but it may be worth noting that the evidence for it is of two main types, neither of which is scientific, so that both come hardly to those trained primarily in science. One type is general and philosophical, and concerned with such problems as morals, the first cause and natural law, while the other is particular and historical, and concerned especially with the events recorded in the New Testament. The nature of the universe and of man presents an extraordinary problem, however it be interpreted, while deductions from general principles could hardly have led men to the kind of conclusions about God's nature and actions which Christians claim were revealed by the life of Jesus Christ. Many will therefore agree with Jeremy Taylor, that the Christian truths 'are articles of so mysterious a philosophy, that we could have inferred them from no premises, discoursed them upon the stock of no natural or scientific principles; nothing but God and God's Spirit could have taught them to us' (*Of the Spirit of Grace*, Sermon, 1651).

Under the impact of Darwinism, the Victorians formed a Metaphysical Society, in which eminent atheists, agnostics, humanists, Anglicans and Roman Catholics, including T. H. Huxley, Tyndall, Tennyson, Ruskin, Gladstone and Manning, met for discussions. They did not convert each other; the evidence both for and against Christianity is not convincing in the same sense that good scientific evidence is convincing. Nevertheless, Christians have not infrequently become atheists and atheists Christians. Either view involves unexplained gaps and contradictions, which relate not to trivial but to fundamental points. Hence the decision is not appreciably easier for the few with great knowledge of science and philosophy than for the huge majority of mankind with scarcely any knowledge of either. All men are in the dark, though Christians would quote St. John's

Gospel (1 : 5) that 'the light shineth in darkness; and the darkness comprehended it not.'

Because the rational arguments for Christianity are by no means compelling, it seems necessary to add personal religious experience, which to some is totally convincing and by others is unfelt or deemed an illusion. St. Paul and many later converts to the Christian faith felt that they did not choose, but rather were called; and on the other side, various modern writers ascribe religious beliefs to irrational but natural causes, such as a transferred father-love or a bourgeois upbringing. For its adherents, the acceptance of Christianity is an act of faith; and faith, J. H. Newman said in a sermon of 1839, 'is an instrument of knowledge and action. . . distinct from those which nature supplies, and. . . independent of what is commonly understood by reason'. Further, 'faith is a principle of action, and action does not allow time for minute and finished investigations', and while it is 'the chosen instrument connecting heaven and earth', it is 'of a nature to excite the contempt or ridicule of the world' (*Fifteen Sermons preached before the University of Oxford*). Since the matter is thus put beyond the province of reason, Christian and agnostic inevitably disagree, though both may accept Newman's conclusion, that 'half the controversies in this world are verbal ones; and could they be brought to a plain issue, they would be brought to a prompt termination. . . When men understand each other's meaning, they see, for the most part, that controversy is either superflous or hopeless'.

The aim of this book is not to solve man's intellectual and metaphysical difficulties, but the much narrower one of assessing Darwinism in relation to them. In this connection, I think it legitimate to draw the following ten conclusions from the arguments of previous chapters, though they may not be agreed by all readers:

1. Animal evolution is an historical fact, and fossils have been found which link man with ape-like forms, so that there is every reason to hold that man evolved from other animals.

2. Various statements concerning natural history in the first three chapters of Genesis are factually wrong. But these chapters should be regarded as allegorical, or at least as allegorical history, which is probably what their writers intended them to be. This need in no way lessen their spiritual truth, which is concerned with matters that come outside science.

3. Evolution is comprehensible in terms of the natural selection of hereditary variations, and so far as known, does not take place in any other way. The variations are random in relation to the needs of the animal, and the directions of evolution are determined by natural selection.

4. For this and other reasons, the concept of either an internal urge or an external Life Force directing the course of evolution is inadmissible. It is also unnecessary and undesirable to postulate that animal evolution has been helped by supernatural interferences with natural laws.

5. The fear that the course of evolution has been entirely 'fortuitous' or 'random' is due to a misunderstanding, since evolution has proceeded in accordance with natural laws. The alternative fear that its course has been rigidly predetermined by mechanistic forces is likewise due to a misunderstanding, since evolution has taken a particular historical course; and the true nature of scientific laws and of historical sequences, and their connection with causation and determinism, are hard problems in philosophy on which the theory of evolution throws no special light.

6. That the universe appears to be run according to natural laws does not, in itself, provide a compelling argument for either theism or atheism, though such a claim has been made both ways. While, too, some have vividly felt the existence of God from the grandeur of the universe or the beauty of living things, others have felt nothing of the kind.

7. All should accept the findings of science in the field of science. The agnostic T. H. Huxley—'follow humbly wherever nature leads'—and the parson Charles Kingsley—'science is the Voice of God'—speak to the same effect on this point. Hence though it may be hard for some Christians to reconcile natural selection with the God of mercy, or for some secular humanists to reconcile man's evolution by natural selection with morality or beauty, such difficulties provide no valid reason for doubting scientific evidence.

8. On the other hand, it is important that the claims made by scientists in the name of science should relate to genuinely scientific matters, and that when they really refer to philosophical problems, this should be made clear. In particular, the claim that man has evolved wholly by natural means is philosophical and not scientific.

9. The theory that man's moral behaviour has been evolved, directly or indirectly, by natural selection fails to account for the essential aspects of the moral experience. Yet no other means of evolution is admitted by biologists.

10. Science has not accounted for morality, truth, beauty, individual responsibility or self-awareness, and many people hold that, from its nature, it can never do so, in which case a valid and central part of human experience lies outside science. But if man evolved wholly by natural means, it might be supposed that all human nature should be interpretable in scientific terms. It might therefore be argued

that man cannot have evolved wholly by natural means. But others would disagree, since there are unbridged gaps and unreconciled contradictions in every view of the meaning, or lack of meaning, of the universe.

These ten points summarize what may reasonably be concluded from the evidence, but they do not take us nearly far enough for deciding the basis for our lives and conduct. Hence the question is, which set of gaps and contradictions are to be accepted. A Christian, agreeing to man's evolution by natural selection, has to add that man has spiritual attributes, good and evil, that are not a result of this evolution, but are of supernatural origin. A secular humanist, likewise agreeing to evolution by natural selection, accepts the validity of morality, truth and beauty, while acknowledging that their genesis cannot yet be established. At the present day, both these views are honestly held, and one need not doubt the integrity of either Christian or agnostic biologists. On the other hand, it is extremely hard to maintain that moral and other values are, or could be, a product of biological evolution, or alternatively, that because they are inexplicable in terms of science, their value is illusory and should be disregarded.

At this point in the book, one critic commented that I had now got the two chief combatants satisfactorily into the ring, stripped clear of inessentials so that the real fight could begin, and that then, just as the climax seemed imminent, I disappointingly left off. But this is to misunderstand my aim, which, as already mentioned, is not to plead for theism or atheism, but to assess Darwinism in relation to them. The problems discussed in this book, in so far as they are scientific, have been largely solved, while in so far as they are philosophical, the theory of evolution and its consequences seem to have advanced them little further. Readers may be disappointed that Darwinism has not achieved more, especially after the high hopes once held out for it. Yet it is not really surprising that it has failed to solve the ultimate problems of philosophy; and a worth-while advance has been made, if only in a better understanding of the limitations of human inquiry.

If, with T. H. Huxley, we 'follow humbly wherever nature leads', we must acknowledge that the essence of human nature remains as mysterious and exciting as before. It therefore seems fitting to end with some selected couplets from Alexander Pope's *Essay on Man* of 1732–4, though similar views could be found in much earlier and later writers:

Placed on this isthmus of a middle state,
A being darkly wise and rudely great;
He hangs between; in doubt to act or rest;
In doubt to deem himself a god, or beast;
Created half to rise, and half to fall;
Great lord of all things, yet a prey to all;
Sole judge of truth, in endless error hurled;
The glory, jest and riddle of the world.

Appendix I

Page references to quotations from books cited in the text, listed in the order cited

NOTES

(i) When more than one quotation from the same book occurs in the same chapter, the title is not repeated, but the page numbers concerned are given in the order in which they are quoted in my text.

(ii) When the quotation has been taken from a later edition than the first, the date of publication has been placed in brackets (the original date is given in the text).

Frontispiece

CHARLES DARWIN, *The Life and Letters, ed.* F. Darwin, 1887, 1 : 313

Preface

BROWNE, SIR T. (1927), *Works, ed.* C. Sayle, *Religio Medici*, pt. 1, sect. 6, p. 12

Chapter 1

VERNON, H. M. and D., 1909, *A History of the Oxford Museum*, 59

HUXLEY, L. 1903. *Life and Letters of Thomas Henry Huxley*, 1 : 266, 217, 316, 368

CLARKE, C. P. S. 1932. *The Oxford Movement and After*, 81

WILBERFORCE, S., in *Famous Reviews, ed.* R. B. Johnson (1914), 279–80 (review of *The Origin of Species* from *Quarterly Review* for July, 1860)

DISRAELI, B. 1864. *Church Policy*, 26

HORT, A. F. 1896. *Life and Letters of Fenton John Anthony Hort*, 1 : 414, 374

KINGSLEY, C. *Life, ed.* by his wife, 1877, 2 : 171

NEWMAN, J. H. (1913). *Apologia pro Vita sua*, 353–4

HEDLEY, BISHOP, *ed.* C. Butler (1931). *Evolution and Faith*, 54 (repr. from *Dublin Rev.*, 1871)

MIVART, ST. G. 1871. *On the Genesis of Species*, 277

FARRAR, F. W. 1886. *History of Interpretation*, 428

Chapter 3

LIDDON, H. P. 1897. *Life of Edward Bouverie Pusey*, 4 : 54

TEMPLE, W. 1934. *Nature, Man and God*, 309

SUTCLIFFE, E. F. 1953. *A Catholic Commentary on Holy Scripture* (*ed.* B. Orchard *et al.*), paras. 142a, 138d.

GLADSTONE, W. E. 1886. Proem to Genesis, *Nineteenth Century*, 107 : 14, 17

BROWNE, SIR T. (1927). *Works, ed.* C. Sayle, *Religio Medici* pt. 1 sect. 45 p. 65.

PIUS XII. 1950 (trs. R. A. Knox, 1952) *Humani generis. False trends in modern teaching*, para. 36, 38, 37, pp. 21–23

Chapter 5

HERSCHEL, J. 1861, cited in *More Letters of Charles Darwin, ed.* F. Darwin and A. C. Seward, 1903, 1 : 191

MIVART, ST. G. 1871. *On the Genesis of Species*, 242

BUTLER, S. (1913). *Evolution Old and New*, 3

RAVEN, C. E. 1953. *Natural Religion and Christian Theology*, 2 : 134, 137, 139

RAVEN, C. E. 1943. *Science, Religion and the Future*, 108

SMETHURST, A. F. 1955. *Modern Science and Christian Beliefs*, 123, 120–1

BERGSON, H. (trs. A. Mitchell). 1911. *Creative Evolution*, 72

FISHER, R. A. 1954. Retrospect of the criticisms of the theory of natural selection in *Evolution as a Process*, *ed.* J. Huxley, A. C. Hardy and E. B. Ford, 91

Chapter 6

PALEY, W. (1845). *Natural Theology*, Bk. 2, Ch. 10. pp. 143–4

TEMPLE, F. 1884. *The Relations between Religion and Science*, 115

TEMPLE, F. 1860. *The Present Relations of Science to Religion*, 6

DRUMMOND, H. 1894. *The Ascent of Man*, 426

KINGSLEY, C. *Life*, (*ed.* his wife) 1877, 2 : 171

COULSON, C. A. 1955. *Science and Religion, a changing relationship*, 2

SHAW, B. 1921. *Back to Methusaleh*, Preface xl.

SMETHURST, A. F. 1955. *Modern Science and Christian Beliefs*, 118

HUXLEY, T. H. 1887. On the reception of the *Origin of Species*, being Ch. V in vol. 2 of *The Life and Letters of Charles Darwin*, *ed.* F. Darwin, 2 : 199, 202

FISHER, R. A. 1950. *Creative aspects of natural law*, 15, 16

KINGSLEY, C. *Life*, *ed.* by his wife (abr. ed. 1883), 318 (*cf.* 1877, ed. 2 : 373)

THOMSON, J. 1880. *The City of Dreadful Night*, 38

TENNYSON, A. (1913). *Works*, *ed.* Hallam Lord Tennyson. The Day-dream, Moral ii, p. 107; In Memoriam cxxiv p. 282

COULSON, C. A. 1955. *Science and Christian Belief*, 40
KINGSLEY, C. 1880. *Scientific Lectures and Essays*, 327
HUXLEY, L. 1903. *Life and Letters of Thomas Henry Huxley*, 1 : 314
NEWMAN, J. H. (1900) (*new imp.*) *Fifteen Sermons preached before the University of Oxford* (Sermon x) 194

Chapter 7

PALEY, W. (1845). *Natural Theology*, Bk. 3, ch. 24, p. 144
TENNYSON, A. (1913). *Works ed.* Hallam Lord Tennyson, *In Memoriam*, LV, LVI, p. 216
SHAW, B. 1921. *Back to Methusaleh*, Preface xl.
RAVEN, C. E. 1943. *Science, Religion and the Future*, 104
HOOKER, R. (1908) *Of the Laws of Ecclesiastical Polity*, Bk. 3, xi, 21. *Everyman ed.* R. Bayne, 1 : 358
PUSEY, E. B. 1878. *Un-science, not Science, adverse to Faith*, Notes, p. 52

Chapter 8

HUXLEY, L. 1903. *Life and Letters of Thomas Henry Huxley*, 1 : 246, 2 : 16
DARWIN, C. (1890). *The Descent of Man*, 613, 126
LIDDON, H. P. 1882. *The Recovery of St. Thomas*, with a prefatory note on the late Mr. Darwin, 13

Chapter 9

SPENCER, H. 1879. *The Data of Ethics*, 28, 18, 25
MIVART, ST. G. 1889. *The Origin of the Human Reason*, 4, citing R. Lankester in *Encycl. Brit.*, 24 : 820
DARWIN, C. (1890), *The Descent of Man*, 130

HUXLEY, J. in Huxley, T. H. and J. (1947), *Evolution and Ethics*, 124
HUXLEY, J. 1953. *Evolution in Action*, 146
SIMPSON, G. G. (rev. abr. 1951) *The Meaning of Evolution*, 156, 159
WADDINGTON, C. H. 1942. *Science and Ethics*, 17, 18
MATTHEWS, W. R. in Waddington, C. H., 1942. *Science and Ethics*, 23
HALDANE, J. B. S. 1928. *Science and Ethics*, 32
HUXLEY, L. 1903. *Life and Letters of Thomas Henry Huxley*, 3 : 220
HUXLEY, T. H. in Huxley, T. H. and J. (1947) *Evolution and Ethics*, 80, 82
BALFOUR, A. J. 1895. *The Foundations of Belief*, 83

Chapter 10

DARWIN, C. *The Life and Letters*, *ed.* F. Darwin, 1887 1 : 316
SHAW, B. 1921. *Back to Methusaleh*, Preface xl.
COULSON, C. A. 1955. *Science and Christian Belief*, 89
PATON, H. J. 1955. *The Modern Predicament*, 324
KNOX, W. L. and VIDLER, A. R. 1933. *The Development of Modern Catholicism*, 129
NEWMAN, J. H. (1913). *Apologia pro Vita sua*, 335
PIUS XII. (trs. R. A. Knox, 1952), *Humani generis. False trends in Modern Teaching*, para. 37, p. 22
SIMPSON, G. G. (rev. abr., 1951) *The Meaning of Evolution*, 155
HUXLEY, L. 1903. *Life and Letters of Thomas Henry Huxley*, 3 : 220
TAYLOR, F. S. 1951. *Man and Matter*, 187

APPENDIX

Chapter 11

TAYLOR, J. *Famous English Sermons, ed.* A. Sampson (1940) 132

NEWMAN, J. H. (1900) (new imp.) *Fifteen Sermons preached before the University of Oxford* (Sermon x) 179, 188, 177–8, 201

Index